Change Yo [obscured]

“A fascinating look into [obscured] natural circadian rhythms. Dr. [obscured] us insight into the Ayurvedic wisdom regarding healthy daily routine and gives practical tips on aligning ourselves with these rhythms for optimal health.”

—Dean Ornish, MD, founder and president, Preventive Medicine Research Institute and author of *The Spectrum* and *Dr. Dean Ornish’s Program for Reversing Heart Disease*

“Look and feel younger by resetting your body’s internal clock. Dr. Kshirsagar’s new book will give you a powerful and actionable Ayurvedic template to lead a mindful and active life.”

—Sara Gottfried, MD, *New York Times* bestselling author of *Younger, The Hormone Reset Diet*, and *The Hormone Cure*

“More and more, we are learning that our best prescription for healthy aging and effectively staving off disease depends on good habits in the areas of sleep, diet, exercise, and managing stress. This book covers it all! And it’s all about the choices you make every day!”

—Dr. Rudolph E. Tanzi, professor of neurology at Harvard Medical School, and bestselling author of *Super Brain, Super Genes*

“The latest research in chronobiology clearly shows how our physical, mental, and emotional health are affected by natural rhythms. The timing of meals, exercise, and sleep is more important than your work schedule. I advise people to follow the cycles of nature for a healthy and happy life.”

—Andrew Weil, MD, author of *8 Weeks to Optimum Health* and *Healthy Aging*

"If you've struggled with your health, you may not be doing the wrong things, you may be doing the right things at the wrong times. Dr. Kshirsagar's book unlocks an easy to use blend of ancient wisdom and modern science. He will teach you how to live a life based on your body's schedule so that you can effortlessly regain sleep, shed weight, and regain energy."

—Alan Christianson, NMD, *New York Times* bestselling author of *The Adrenal Reset Diet*

"Dr. Kshirsagar's book and its deep knowledge about chronobiology as it affects everyone's daily schedule looks to a future that advances the evolution of self-care."

—Deepak Chopra, MD

"A brilliant book full of expertise and practical advice, *Change Your Schedule, Change Your Life* provides a fantastic and easy to read roadmap to your healthiest life possible."

—Marci Shimoff, #1 *New York Times* bestselling author of *Happy for No Reason* and *Chicken Soup for the Woman's Soul*

"Circadian medicine is destined to revolutionize medicine as we know it. Living recklessly out of sync with nature's cycles is unsustainable and linked to most of the chronic and degenerative diseases we face. *Change Your Schedule, Change Your Life* makes a promise it will keep while adding years of health, joy, and longevity to your life."

—Dr. John Douillard, DC, CAP, bestselling author of *Eat Wheat* and *The 3-Season Diet*, former NBA nutritionist, and the creator of LifeSpa.com

"One of the simplest and most effortless ways to improve your mental, emotional, and physical health is by living in tune with the Ayurvedic schedule. It's the easiest path to a happier and healthier life—living in alignment with your natural rhythms."

—Dr. Kulreet Chaudhary, author of *The Prime: Prepare and Repair Your Body for Spontaneous Weight Loss*

"Dr. Suhas Kshirsagar has an extraordinary ability to share the ancient wisdom of Ayurveda with clarity, inspiration, and humor. This book is a must read for anyone who is struggling with their busy lifestyle, busy schedule, and/or who wants to understand more about a holistic approach to everyday life, health, and balance."

—Daniel Vicario, MD, ABIHM, medical director and director, Integrative Oncology Program San Diego Cancer Research Institute

"Living in tune with the rhythms of life is one of the key solutions to our modern ills. Dr. Kshirsagar combines the ancient wisdom of Ayurveda with modern research to help you get back in touch with these rhythms and create the optimal daily schedule that is best for you—and can improve every aspect of your life."

—Akil Palanisamy, MD, author of *The Paleovedic Diet*

"Dr. Kshirsagar addresses the root cause of the issues and expertly guides us to the solutions that lead to our success. This is rare and priceless. Following this advice will liberate you from being a slave to your schedule and always being the victim of low energy and no focus. It will help you live again."

—Yogi Cameron, author of *The Yogi Code*

"The latest science of chronobiology validates the importance of bodily rhythms. There is no substitute for healthy diet, good sleep, and a positive lifestyle. Read the book."

—Dr. Murali Doraiswamy, professor,
Duke University Health System

"With his book *Change Your Schedule, Change Your Life*, Dr. Suhas reveals the benefits of living a life aligned with the body's natural rhythms. Grounded in the ancient science of Ayurveda, Dr. Suhas gives the reader the tools necessary to make small, manageable changes to truly transform their quality of life. Dr. Suhas cuts through the chatter of fad diets and quick fixes and offers real suggestions healthy, simple, manageable change."

—Anand Dhruva, MD, associate professor of medicine,
University of California San Francisco

Change Your Schedule, Change Your Life

ALSO BY DR. SUHAS G. KSHIRSAGAR

The Hot Belly Diet

Change Your Schedule, Change Your Life

◒ ○ ◑ ●

How to Harness the Power of Clock Genes to Lose Weight, Optimize Your Workout, and Finally Get a Good Night's Sleep

◒ ○ ◑ ●

Dr. Suhas G. Kshirsagar

WITH MICHELLE D. SEATON

Foreword by Deepak Chopra, MD

HARPER WAVE
An Imprint of HarperCollins*Publishers*

A hardcover edition of this book was published in 2018 by Harper Wave, an imprint of HarperCollins Publishers.

This book contains advice and information relating to health care. It should be used to supplement rather than replace the advice of your doctor or another trained health professional. If you know or suspect you have a health problem, it is recommended that you seek your physician's advice before embarking on any medical program or treatment. All efforts have been made to assure the accuracy of the information contained in this book as of the date of publication. This publisher and the author disclaim liability for any medical outcomes that may occur as a result of applying the methods suggested in this book.

FIRST HARPER WAVE PAPERBACK EDITION PUBLISHED 2019.

The Library of Congress has catalogued the hardcover edition as follows:

Names: Kshirsagar, Suhas G., author.

Title: Change your schedule, change your life : how to harness the power of clock genes to lose weight, optimize your workout, and finally get a good night's sleep / Dr. Suhas Kshirsagar with Michelle D. Seaton ; foreword by Deepak Chopra, MD.

Description: First edition. | New York, NY : Harper Wave, [2018] | Includes bibliographical references.

Identifiers: LCCN 2017053533 (print) | LCCN 2017050677 (ebook) | ISBN 9780062684882 (E-book) | ISBN 9780062684851 (hardback) | ISBN 9780062684868 (paperback) | ISBN 9780062684851 (digital edition)

Subjects: LCSH: Circadian rhythms—Health aspects—Popular works. | Medicine, Ayurvedic. | Health behavior. | BISAC: HEALTH & FITNESS / Weight Loss. | HEALTH & FITNESS / Sleep & Sleep Disorders. | HEALTH & FITNESS / Healthy Living.

Classification: LCC QP84.6 (print) | LCC QP84.6 .K74 2018 (ebook) | DDC 612/.022—dc23

LC record available at https://lccn.loc.gov/2017053533

ISBN 978-0-06-268486-8 (pbk.)

23 24 25 26 27 LBC 12 11 10 9 8

Contents

Foreword

My father was proud of being a physician trained in the scientific method so, not surprisingly, I became a doctor in the same mold. It took a process of personal evolution for me to recognize how valuable Ayurveda actually is, and how compatible it is with other approaches in the current wellness movement. Today, Ayurveda can realistically be called a pillar of integrative or whole system medicine.

When I wrote *Perfect Health* in 1991, which laid out the principles of Ayurveda for everyday life, I wondered if readers would embrace the appeal of making lifestyle choices very different from the Western default pattern. But I was encouraged by the natural interest people took in discovering their body type, the basic entry point into Ayurveda, which then leads to personalized diet and seasonal routines. Even more important, *Perfect Health* focused on consciousness as the most powerful agent for changing body and mind. A consciousness-based Ayurveda goes far beyond the notion of Ayurveda as "alternative medicine." Instead, it is elevated to being about a person's evolution in every dimension: physical, mental, and spiritual. In the legends of Ayurveda, there are intense practices that supposedly created immortals; in real Ayurveda, you realize your status as immortal to begin with, defeating the illusion of birth and death.

But medicine in the West—and increasingly in India, China, and the East in general—hasn't focused on expanding awareness. Quite the opposite! The ideal has been to devise a kind of safety net based on eating the right foods, getting the right amount of exercise, managing stress, and controlling the various negative influences, like smoking and alcohol, that damage people's health and shorten life expectancy. Society has reached a plateau in this regard, I think, because the notion of avoiding risk is anxiety-based. Wellness becomes an insecure state that is bound to be temporary given the many external assaults from the immediate environment.

Ayurveda doesn't contradict these measures for achieving wellness, but the primary focus is holistic balance, which leads to a profound trust in Nature, beginning with your own body's connection to the environment. The entire Indian wisdom tradition comes down to ending separation and living in unity consciousness. It isn't that unity is a prize held out after a lifetime of arduous practice. Instead, unity is the ground state of existence, from which we have become separated. Returning to the ground state, or being authentic, must involve a natural way of life that keeps the bodymind in balance while also evolving—in other words, the development of the domain of consciousness, of the world "in here," is all-important.

No system of "alternative medicine" can be expected to achieve unity consciousness. It is closer to the truth to use the Sanskrit term *Upaveda*, where *veda* means "the teaching about reality" and *upa* means "close to or near." An Upaveda isn't pure spiritual teaching, but an adjunct or helper that's close to the pure teaching. In the West, this seems like a dubious role for medicine, because scientific medicine is essentially the equivalent of taking a car to a mechanic to have it repaired. In fact, the mechanistic approach taught in medical school is worn as a badge of honor: a good doctor ignores the changing, untrustworthy world of a patient's inner feelings, thoughts, habits, inclinations, or anything else that is considered subjective. Even psychiatry, which is the one specialty that crosses the boundary into a patient's inner world, has become largely a matter of matching symptoms with the appropriate drug—all the time knowing that drug treatments are rarely if ever a cure for the underlying mental disorder.

When they aren't visiting the doctor, people in everyday life spend very little time examining the lifestyle they grew up with, much less aiming for the Ayurvedic ideal, which is to be aware of shifting conditions and mind and body on a daily basis. Such awareness, in the sense of mindfulness, isn't the same as being anxious about what you're eating and how you feel. Taking the "upa" part of Upaveda seriously, the routine you follow every day and every season of the year is a helper in achieving a higher state of well-being on all fronts.

Coming to the focus of this book, Western medicine has been undergoing its own quiet revolution with the rise of chronobiology, the study of how time affects the physiology in gross and subtle ways. As mounting evidence has proven, timing is everything inside the body. Every process in trillions of cells is regulated by an internal clock—one that turns out to be very similar to the one described in the Vedic texts. In fact, the critical importance of daily (circadian) rhythm may prove to be the link between ancient Ayurvedic practices and alleviating the modern epidemic of chronic diseases.

In 2017, three research physiologists won the Nobel Prize for their four decades of work unraveling the mysteries of the circadian rhythm in biology. They found that the diurnal rhythm of nature affects the functioning of cells in plants, animals, humans, and even some single-cell bacteria. Specific genes change the function of cells based on the time of day. While this may seem like an esoteric set of discoveries, the new field of chronobiology has practical applications that have revolutionary implications for the future of wellness.

It is well established by now that lifestyle choices can change the expression of our DNA, but what we have been learning recently is that it's not enough to eat well, exercise a few times a week, and get sound sleep. As Ayurveda has taught for centuries, you must know which daily schedule works *with* your physiology and not against it.

This realization is what makes *Change Your Schedule, Change Your Life* such a valuable addition to the growing awareness of Ayurveda in the West. Despite all the best prevention advice, none of it disputed, millions of people routinely work too many hours and sleep irregularly with their cell phones next to their beds. They eat in a rush, even when not indulging in the national craving for fast food. "Time sickness" creeps into their daily activity, by which is meant living with one eye on the clock, constantly aware of deadlines and an overloaded list of duties and demands.

These unrealistic lifestyle expectations have become something acceptable, but new medical research is undermining the assumption that our bodies can adapt to the abnormal. Chronic imbalance has become a common situation affecting every cell, and the chief culprits are chronic stress and low-level inflammation. If the hunches of leading researchers are borne out, it could be true that literally every lifestyle disorder, including heart disease, obesity, hypertension, and type 2 diabetes, has its roots years and decades before symptoms appear. These roots are the imbalance caused at a subtle level by daily stress, which we take for granted, and chronic inflammation, which is so hidden few people would ever notice it.

The Ayurvedic prescription for a state of imbalance, which applies to both stress and inflammation, is to restore balance and then let the bodymind's natural preference for remaining in balance do the rest. In practical terms, we need to be moving and nourishing and resting our bodies in sync with nature's rhythm. Once we do this, we find it easier to go to sleep at night and get up in the morning, to maintain a healthy weight, and to resist tempting but unhealthy foods. It's even easier to unplug from distractions and find more time for our personal goals.

Ayurveda has been teaching for millennia that there is a link between mind and body, founded on the unity of every natural process. Today Dr. Suhas Kshirsagar leads the next wave of Ayurveda in the West. His book, and its deep knowledge about chronobiology as it affects everyone's daily schedule, looks to a future where self-care becomes far more important than relying on a doctor to make repairs after the symptoms of damage are apparent.

When self-care is based on self-awareness, we approach the Ayurveda ideal set forth by the ancient rishis. Figures like Dr. Kshirsagar are keeping the ideal alive and, more important, advancing the evolution of self-care just when it is most needed. I welcome his book and him as an *Upaguru*, a teacher who sits close to his students and guides them with intimate, caring love and compassion.

—Deepak Chopra, MD

1 It's Not You; It's Your Schedule

Tell me your daily routine, and I'll tell you how healthy you feel. Tell me when you eat, and I'll tell you if it is easy or difficult for you to maintain your weight. Tell me when you exercise, and I'll tell you whether you are building your body's systems or wearing them down. Tell me when you turn off your television or computer at night, and I'll tell you how sensitive you are to stress. Tell me when you fall asleep, and I'll tell you whether you need coffee to power your way through the afternoon, or whether you snap at your loved ones at the end of a long day when you wanted to be patient.

Does this sound like magic? It's not. A growing body of science reveals how closely our bodies are linked to the circadian rhythm of light and darkness, right down to the cellular level. This research shows that when you eat is as important as what you eat, when you fall asleep is as important as how much sleep you get, and when you exercise is as important as how much exercise you get. Your daily schedule determines your weight, your stamina, your general health, and your mood. Don't believe me? For decades, diabetes researchers have known that a simple way to trigger obesity in laboratory mice is to wake them up and feed them during their sleep cycle. In fact, mice gain weight within a week if researchers just expose them to low-level lights when they should be sleeping.[1]

Still don't believe me? Think back to the last time you experienced jet lag. How did you feel? Anyone who has experienced jet lag knows the symptoms can go far beyond sleep disruption. Often, you suffer from constipation, upset stomach, cognitive fog, low energy, and an increased sensitivity to stress. A recent study even linked jet lag to weight gain because disrupting your schedule through long-distance travel confuses the microbes in your gut.[2]

Yet these same complaints—weight gain, insomnia, exhaustion, stress, depression—are the very things that bring people to my clinic. And, if you're reading this book, I'd guess that these complaints sound awfully familiar to you, too. Thanks to the demands of modern jobs and 24/7 connectivity, many of us live in a constant state of self-imposed jet lag, sleeping, eating, and exercising at times that don't coincide with the body's natural rhythms. But there is good news, and I'm here to tell you what I tell all my patients: *It's not you; it's your schedule.* There is an easier way to lose weight, get energized, and get to sleep at night. By working with your body's natural rhythms and not against them, you can create a daily schedule that will transform your health and your life.

The Circadian Rhythm

Physiologists know that the body has a natural rhythm—called a circadian rhythm—that operates on a nearly twenty-four-hour cycle, resetting itself every morning when you first experience daylight. This rhythm directs the body on when to digest food, how to prepare for sleep, and how to regulate everything in your body including blood pressure, metabolism, hormone production, body temperature, and cellular repair. Your skin cells, too, repair and regenerate on a daily schedule. Even the population of microbes in your intestinal tract changes profoundly throughout the course of a single day. Certain strains of gut bacteria proliferate during the day, while others predominate at night. At every hour of the day, your body is changing its function. The cells and systems are primed to do different things, depending on the time of day or night. That's how we know that you hit your deepest sleep cycle at about two a.m., that your body temperature is lowest at about four a.m. Your body's sharpest rise of blood pressure comes at about six forty-five a.m., and a bowel movement is most likely at eight thirty in the morning. By ten in the morning, your mental alertness peaks, and your digestion is operating most efficiently at noon. Your coordination, reaction time, and cardiovascular strength peak in the afternoon while your digestion powers down. After sunset, your blood pressure hits its highest daily level, along with your body's temperature. At about nine p.m., your brain starts releasing melatonin, and your digestion slows to half speed. By ten thirty, your bowel movements are suppressed, and your digestion is at a crawl. This happens, or should happen, every day. This is why your body gets so confused when you cross time zones. The light changes and the body loses its compass for controlling all of these bodily functions.

This is fascinating because we think we are so isolated from nature. We live in climate-controlled homes and work in offices or cubicles. And yet every system in our bodies is changing in a predictable, daily pattern. Your body is always trying to coordinate all of its systems on a central clock using available natural light. Every organism in nature operates in this cyclical way, and a new field within biology, called chronobiology, studies all the ways in which different organisms operate in accordance with a circadian rhythm.

What researchers are studying now is how our daily habits interact with this circadian rhythm, and they've discovered that the modern schedule profoundly disrupts it. Staying up late at night watching TV or doing work fools your body into thinking that night hasn't started yet. Eating a big meal in the evening does the same thing. It delays the cycle and disrupts sleep, only to have you jolt the body awake first thing in the morning when your alarm goes off. Lack of exercise and natural light further disrupt the circadian rhythm, which in turn disrupts everything from your digestion to your hormone secretion and your nervous system.

Many of my patients routinely stay up until midnight, working and snacking, and then wonder why they can't fall asleep until one. Then they drag themselves out of bed at six a.m. and wonder why they can't eat or concentrate in the morning. A couple of hours of deviation from your body's natural rhythm may not seem like much, but to put this in perspective: if you only sleep between one and six a.m., it's as though you flew from California to New York in the evening, only to fly back before work. No wonder you feel sick.

Many of our most common physical complaints are created or exacerbated by a modern schedule at odds with the body's needs. Fortunately, physiologists have generated a lot of new research about the body's clock and how behavior either helps strengthen the clock's signals or gets in its way. This new field is called chronobiology, and it offers insights about how you can set a daily schedule that will keep you healthy and energized.

How Your Body Tells Time

Your body always knows what time it is, even if you don't. It may sound absurd to think that you don't know what time it is. You are probably hyperaware of the time at every moment of the day. You have a train to catch or you have kids to drop off at school. You have a meeting in fifteen minutes and a call in an hour. You have to get to the dry cleaner before it closes. You have project deadlines, dinner reservations, and an alarm clock (or two) that wakes you every morning. My patients tell me that they are constantly aware of the time and that the clock dictates nearly every one of their daily activities.

But there is a different kind of clock inside your body, one that rules all of its cells and systems. To understand how it works, you have to step inside the brain and into the hypothalamus. The hypothalamus sits at the center of the brain and is responsible for regulating all of the body's systems. It activates the fight-or-flight response when you feel stress or danger. It tells you when you are hungry or thirsty. When you begin a strict diet, the hypothalamus is what's telling you that you are starving because you are eating differently. You may know that you aren't starving, but the body is signaling to the brain that it's not getting the same amount of food as before. When you start a new exercise routine, the body signals muscle fatigue and cardiovascular stress to the brain and the hypothalamus urges you to stop. And when you stay up late to work on a project, the hypothalamus is what's telling you that you are sleepy and bored. So this part of the brain can read the body's signals and try to affect your behavior, trying to keep everything the same as it was yesterday.

The hypothalamus also regulates all kinds of things that you don't consciously control, including body temperature, hormone balance, and metabolism. All of these changes happen at predictable times of day. For example, your body temperature peaks in the evening, then decreases during the night and reaches its lowest point just before dawn. Your blood pressure rises sharply as you wake up each morning, and then increases slowly throughout the day before falling during the night. The sharp rise of blood pressure in the morning comes at a time when blood platelets are stickiest, which explains why many heart attacks happen first thing in the morning. Cortisol levels, too, change at predictable times. Cortisol is a steroid that the body produces and it's sometimes called "the stress hormone." The level of cortisol in your body is lowest when you go to bed and then accumulates during the night. It is partially responsible for your body's inflammatory response, so it's no wonder that those aches and pains are at their worst when you get out of bed, or that you feel most bloated and puffy in the morning. Cortisol levels discharge steadily throughout the day, fluttering briefly upward after every meal.

Colonic motility—which is a fancy term for bowel movements—changes during the day as well. First thing in the morning, the colon wakes and moves at three times its normal level of activity, with predictable results. That's why so many people feel constipated while in the throes of jet lag. A poor eating schedule can also confuse the colon. At night, the colon rests and bowel movements are suppressed. Mood and brain waves alter throughout the day and night as well.

In order to regulate the body's systems, your hypothalamus takes its cues both from the body's tissues and organs and from the environment. When you smell food, you feel hungry; when you see danger, you feel anxious and energized for action. All true. But let's don't forget the most pervasive signal the brain takes in all day—the presence of light.

There is a small part of the hypothalamus, called the suprachiasmatic nucleus (SCN) that is tasked with noticing light. It's about the size of a grain of rice, and it contains approximately twenty thousand neurons. Physiologists have long understood that these neurons respond to light and regulate the body's systems based on light and darkness. When light hits the retina of the eye first thing in the morning, the SCN signals to the body that it's daytime. In the evening, the SCN helps signal the body's natural production of melatonin that tells you when it's time for sleep. But it's only in the past twenty years that researchers have looked at how much power this tiny bundle of neurons exerts over every cell and system in the body.

A Brief History of Chronobiology

In order to appreciate the field of chronobiology, we have to travel back almost three hundred years, to an experiment carried out by the French scientist Jean-Jacques d'Ortous de Mairan. In 1729, de Mairan became interested in the way that some plants open their leaves in the sunlight and close at night. So he exposed these plants to constant darkness and observed them. They continued to open their leaves in the morning and close them at night even though it was dark all the time. Their leaves moved as though expecting the sunlight that never came. De Mairan was baffled, and so were the many researchers who replicated his experiment. Another scientist referred to the closing of the leaves as a kind of "plant sleep." The plants continued to open and close their leaves on schedule for many days after sunlight had been blocked. De Mairan wondered if the plants could somehow sense the sunlight aboveground. He didn't go so far as to suggest that the plants had a cellular predisposition to open their leaves at a certain time—it would have been heresy to suggest such a thing, and it remained a kind of heresy for about two hundred years. Instead, de Mairan wondered if ambient changes in temperature or the rotation of the earth was informing the behavior of these plants.

A bigger mystery was why the natural rhythm of opening and closing didn't follow a twenty-four-hour period. Eventually, when scientists could study the plants more closely, they found that these movements became less pronounced in total darkness, and the plants opened and closed their leaves on a twenty-two-hour cycle. But when the plants could experience light, they reverted to a twenty-four-hour cycle. This suggested that they somehow have a biological predisposition to move in anticipation of light, and that the light itself helps them synchronize their internal clocks. It was easy to theorize about how light and darkness affect plants because they need light, but it would take a particular kind of scientist to notice that other types of organisms, including mammals, also use light to alter their physiological function.

That scientist turned out to be a young Romanian doctor named Franz Halberg, who was completing a fellowship at Harvard in the late 1940s when he began to track the levels of circulating white blood cells in mice. He continued his research at the University of Minnesota, where he noticed that the white blood cell count peaked during the day and fell at night. Different strains of mice had different levels of circulating white blood cells, but every type of mouse showed the same pattern of sharp rise during the day and a similar fall at night. Soon, Halberg was tracking hourly fluctuations in blood pressure and heart rate in the mice, along with body temperature, and found that these physiological responses varied on a similar twenty-four-hour schedule. By 1959, he had coined the term "circadian rhythm" to account for these changes. Over the ensuing decades, he theorized about and then proved similar variations in humans.[3]

Halberg found that a host of physiological processes, including body temperature, hormone production, blood cell counts, blood pressure and heart rate, glycogen levels in the liver, even cell division, all vary along predictable patterns, all of which seemed to him to be light-dependent. But genetic research was still in its infancy, and few researchers wanted to believe that the body contained an internal clock that varied with time of day or with seasons of the year.

Halberg was certain that fluctuations within these patterns were possible markers for disease. He believed that monitoring blood pressure constantly gave a better prediction for heart attack and stroke than taking a single measure in a doctor's office. That's why he monitored his own blood pressure every thirty minutes each day for the last fifteen years of his life. Perhaps he was on to something. He lived to be ninety-four.

He further theorized that anticancer treatments would be most effective when the core temperature of tumors was highest. He believed that the body operated entirely on a circadian rhythm, and that nutritionists and medical doctors should consider these rhythms as part of any treatment plan. And although chronobiology centers sprang up in major research centers all over the world, it was difficult to prove the efficacy of these theories until the very end of the twentieth century. Halberg himself found it difficult to secure funding for his studies, and to get the subject of chronobiology taught in medical school.

It would be tempting to say that the rest of the medical community dismissed these theories. But the truth is that at the time it was expensive to constantly monitor blood pressure, blood counts, glucose uptake in the liver, and other physiological responses. Halberg's theories ranged far ahead of technology, and it would be up to geneticists to explore these ideas and figure out exactly how the cells in the body are able to coordinate with the SCN each day and night.

Clock Genes

We now know that the cells in the body contain what are known as "clock genes." They have specific names, such as per1, per2, per3, which are active at night, or CLOCK and BMAL1 genes, which are active during the day. They operate on a kind of loop. The activity of one of these clock genes inhibits the activity of the other. Your cells are primed to do different things based on the light and dark cycles of the day. And the protein pathways for each of these cells is active or inactive based on the time of day.

Every morning when you open your eyes and see daylight, your SCN is giving the signal to reset the internal clock and sends information to all of the systems of the body and all of the organs and tissues to say that it's daytime again. And this clock sets the automatic physiological changes that must occur on time for the next twenty-four hours in order for the body to function. In this way, the SCN is the brain's clock. Or perhaps we should say that it is the brain's master conductor, and all of your body's cells are trying to dance to its beat.

While the master clock in your brain is trying to set a total body rhythm, the cells in your body are reacting to your behavior—your sleep schedule, your mealtimes, and your activity—to set their own clock rhythms. When the brain's clock and these cellular clocks—called "peripheral clocks"—are out of alignment, you can get distorted cell behaviors. Remember those mice who gained weight when they were fed during their sleep cycle. Their bodies were operating outside the master circadian rhythm, taking in nutrients that the cells in their digestive tract couldn't process. And the reduced sleep schedule meant that on a cellular level, whole systems in their bodies stopped functioning the way they were supposed to. This causes disruptions not just in the digestive process, but in hormone production, immune response, and inflammatory response.

As you can imagine, this puts an entirely new spin on the field of epigenetics and how our behavior alters and affects genetic expression over time. As a field, chronobiology is still new, but clock genes seem to have an effect on aging and tumor suppression in addition to metabolism. Eating and sleeping at the wrong time disrupts the circadian rhythm and interferes with a healthy metabolism and a strong immune response. Though scientists are still working out many of the nuances and clinical treatment applications, what we know for certain is that you can use your daily schedule to reinforce the circadian rhythm and achieve better health.

Even without the presence of light each morning, your body would try to function on that same twenty-four-hour schedule. Starting in the 1970s, researchers did experiments with people who agreed to live in isolation without daily exposure to natural light. (Unlike de Mairan's plants, they agreed to go live in a cave.) A number of findings have emerged from these experiments over many decades. First, the body's clock will drift without the daily reset of natural light and darkness. The primary means by which the body sets its circadian rhythm is through light. Second, the body can use social cues, such as timing of meals, sleep, and exercise, as a substitute when light signals are absent. All of the body's attempts to synchronize its systems onto a single master circadian rhythm are called entrainment. The body relies on signals to help it reset the circadian rhythm and keep itself functioning optimally. While it prefers to use light and darkness as the primary signal, it can and does use other cues, including your behavior. Everything you do all day is either helping the master circadian clock to synchronize the body's functions, or it's getting in the way. Setting a daily schedule that reinforces your body's natural rhythm is the most powerful health habit you can adopt.

It's not just the human body that tries to set this daily rhythm. All of nature follows this same diurnal pattern. De Mairan's plants were opening their leaves in anticipation of light, even though the light never came to them. Their cells kept trying to keep the daily rhythm. Many types of cells are tasked with doing one set of things in the daytime and another at night. This is true for mammals, for plants, and even for the smallest single-cell bacteria. In the past thirty years, genetic research and microbiology have transformed the study of these natural rhythms. Scientists have been working to discover clock genes inside of cells and how they work at the neurological and molecular level in all types of organisms.

Knowing that cells operate differently at different times in a twenty-four-hour cycle has implications on many different fields. The study of the circadian rhythm and chronobiology could change many types of medical therapies. For example, if you take a short-acting statin to control your cholesterol levels, your doctor is probably going to tell you to take it at night. Why? Because chronopharmacologists know that that is when the liver produces cholesterol. Researchers are searching for the limit of the ways in which circadian rhythms rule biological systems in all living things, but they haven't found it yet. One researcher noted that it should be assumed that every system in the body works on a circadian rhythm until proven otherwise.

Ayurveda and Chronobiology

Although these findings in chronobiology-based research are still new, they actually reinforce what I've been practicing in Ayurvedic medicine for decades. Ayurveda is a natural healing tradition that has been practiced in India for about five thousand years. Long before de Mairan wondered about his plants and their strange behavior, Ayurvedic doctors were cautioning patients about the daily cycle of the body and its many systems. Ayurveda separates the day into segments that describe the body's energies and systems as active or dormant. It teaches that you must have a healthy schedule in order to thrive. In fact, Ayurveda stresses that all of your behaviors, including diet, rest, and exercise, must work together with your master internal clock to keep the body functioning well. It also teaches how to achieve a mind-body connection so that you can stay in touch with what your body needs throughout the day.

In fact, Ayurveda is sometimes called the original lifestyle medicine. Literally translating to "the science of life," it is the precursor to all of the other healing traditions, including traditional Chinese medicine. As Buddhism spread throughout Asia, its scholars brought Ayurvedic knowledge with them. In traditional Chinese medicine, practitioners teach that balance is essential, that the flow of chi through the body facilitates healing, and that the taste of your food is part of its healing and balancing effect. These ideas were influenced by Ayurveda. Even the Greeks read the Ayurvedic texts, and gleaned some of the wisdom to inform their own ideas about how the body works. *Prana*, the Sanskrit word for the breath of life, became *pneuma* to the Greeks. *Agni*, the heat of metabolism and digestion, became *ignis*. And, somehow, the three *doshas*, or prevailing energies, became three of the four humors or metabolic agents—phlegm, cholera, melancholy. In early Ayurvedic texts, one scholar concluded that blood was a fourth dosha, and the Greeks, apparently, agreed, although later Ayurvedic scholars reverted to three doshas. The Greeks believed that these humors, or tendencies, needed to be balanced, and that a disruption or overabundance of one of these humors was the cause of many diseases, ideas borrowed from Ayurveda.

But few other natural healing traditions explore the effect of natural light on the body. Here Ayurveda stands alone in explaining that the body's systems operate on a daily cycle. It describes a daily routine, a nightly routine, and a seasonal routine to sync your body with the circadian clock. The term chronobiology may be relatively new to Western medicine, but it is an essential part of the Ayurvedic tradition. The texts describe how our bodies are constantly interacting with sunlight and the changing seasons. They talk about how to synchronize your daily routine with this changing natural light. It may be the only medical tradition that talks about how to arrange your daily routine to achieve optimum health throughout the seasons and the decades of your life.

Further, it's one of the only traditions to talk about body type and how this manifests in certain health problems. This is significant because so much diet and health advice assumes that everybody is more or less the same in their need for sleep, exercise, and food. If you look around a room, you can see that your body is not like everyone else's. In Ayurveda, there is no concept of a single diet or exercise routine that can help every person. While everyone needs to understand how to set a good schedule, not everyone needs exactly the same diet or exercise routine within that schedule to get the best results. In this book, I'll describe why you need good sleep and outline all of sleep's surprising benefits, while in the next chapter, I'll help you identify and solve your particular challenges with sleep. I'll do the same with diet and exercise. If you have one body type, you may struggle to lose weight, but never experience insomnia. Someone with a different body type may never have worried about being overweight, and yet headaches and insomnia are a kind of curse. But there are answers to all of these concerns. So, if you've tried exercise, and you couldn't stick with it, you can find out why and discover how to solve this problem. If you've tried diets and they didn't work, it's probably because you didn't find the right diet for you. In Ayurveda we try to balance the whole body and we look at the individual and solve the problems exacerbated by body type. In this way, you can set a schedule that will support your body's central rhythm, but you will also be able to fine-tune your schedule so that dieting, sleeping, and getting fit will become effortless.

But first, I want to tell you how to stop getting in the way of your circadian rhythm. By making the body's clock work for you, you'll reap enormous health benefits right away.

2 Using Your Body's Internal Clock

What do you do all day?

This is the first question I ask people who come to my clinic. Many people don't know what they do in a typical day—and this may be true for you, too. You know when the alarm goes off in the morning, and you know how many hours you spend at work, but you might not know when you eat from day to day. You don't know when you typically fall asleep, even if you do know which television shows you watch at night. It's easy to assume that your work schedule dictates when you eat, how much you rest, and when you exercise. When you do that, you stop giving the body the predictable signals it needs to function efficiently. If you eat lunch at eleven thirty on some days and at one thirty on other days, your body is confused already. If you stay up late on some nights to do work or to work out, then your body doesn't know how or when to prepare for sleep.

In order to understand how your schedule already affects your health, you first need to know what that schedule is. Second, you need to ask two basic questions: 1) Am I getting seven hours of uninterrupted sleep when it's dark outside? 2) Am I doing almost all of my eating when it's light outside? Most people cannot answer yes to either of those questions. And that's a problem.

One of my patients, Rory, is a good example of the impact that schedule has on your overall well-being. A Silicon Valley software developer with small children at home, Rory managed to do all the "right" things: he worked out daily, ate a protein-dense, low-calorie diet with plenty of vegetables, and even made time for family. But despite what he thought were healthy choices in the context of a demanding job, his stomachaches were getting more disruptive and he was having trouble falling asleep before two or two thirty. On those nights he was able to get to sleep, he was waking up around four, feeling anxious and in pain.

When we started talking about not just what he was doing but when he was doing it, the rest of the puzzle fell into place. Rory described a typical day of getting up to leave for work by seven a.m. Not usually hungry in the morning, he would often have coffee to get him through and tide him over to his post-workout protein smoothie at one p.m. He would head home in the late afternoon to spend time with his wife and kids, and they would eat dinner as a family at about eight p.m. Once the kids were in bed, he would make a pot of coffee and stay up until at least midnight—and sometimes much later—to work on a special project in coordination with other developers working in India.

Like most people, Rory had no idea that his schedule was the cause of his health problems. If you think food is merely fuel, then you can eat at any time of day. Skipping meals is just saving calories for later. A healthy choice, such as a protein shake, is healthy no matter when you drink it. If exercise is good for you, it must be good no matter when you do it. For many busy and ambitious professionals, sleep is just that thing you do when you haven't scheduled anything else.

This is backward.

Your lifestyle is not the sum total of how many calories you consumed in a single day, how many minutes you logged on a treadmill last week, and how many hours of sleep you typically get. Instead, it's the coordination of these things with the body's scheduled needs. Rory was skipping some meals and replacing others with light fare. He was exercising when his body couldn't benefit from that strenuous activity. And he was eating and working late at night, robbing himself of restful sleep. In the previous chapter, I outlined a chronobiologist's view of the body's circadian rhythm. Now I'll describe Ayurveda's view on this same body clock. You may find them oddly similar.

The Ayurvedic Day

In Ayurveda, the day is separated into six segments that outline the body's needs. The actual texts are quite detailed, but even this generalized overview will give you an understanding of how the body follows circular, diurnal patterns. The descriptions may feel overly poetic or even simplistic at times, but it's important to see that these teachings put the body at the center of a daily cycle, something that chronobiologists confirm with every new study they publish, and something that Rory had never considered.

Six to ten a.m. is predominated by *kapha* energy, which has the qualities we associate with water. This means that the body may be a little bit dull and heavy, prone to retaining water or generating congestion. The mind and body are still waking up with the emergence of daylight, and it needs a jolt of exercise, meditation, and food to synchronize itself with the new day. When balanced with a little bit of exercise and food at this time of the day, the kapha energy stops being dull and instead becomes a steady, calming influence on your morning work.

Ten a.m. to two p.m. is predominated by *pitta* energy, which has qualities we associate with fire. During this time of day both the mind and the digestion are going full steam. It's a good time for your largest meal of the day, and to do your most intense work. Your body doesn't need exercise at this time of the day because you are already wide awake. Moreover, the body needs to keep the blood flow concentrated around your digestive tract so your body can do the work of converting food into energy. This pitta energy also makes people a little more passionate and short-tempered. You might say that your noontime crankiness is all about low blood sugar, but something else is at work here, too.

Two to six p.m. is predominated by *vata* energy, which has qualities we associate with air. It's a time for quick reflexes and quick thinking. This is also a time of day that leads some people to get distracted or dehydrated. If you haven't eaten enough in the morning and at noon, this light and quick energy can become shakiness and anxiety, which is why people reach for snacks and coffee at this point in the day. You want to have grounded yourself with food and exercise earlier in the day, or this lightness will carry you away, making it impossible to concentrate. The body's natural energy comes in fits and starts in the late afternoon, and you may find that you need more rest between tasks. You want to stay hydrated and minimize distractions.

Six to ten p.m. is the time for the energy to switch back to kapha, when the body gets to be a little bit dull and heavy again, preparing for sleep as the sun goes down. As soon as you get to six p.m., the digestion is already slowing down, so this is the wrong time to overload your body with calories. Your mind is moving from a quick and distractible thinking style into something more steady, and some people say they prefer to stay at their desks after six p.m. because they love that feeling of steady energy. But it's easy to overwork the mind, making sleep impossible later, when you need it. Instead, it's best to eat a very light meal early in this period and spend the rest of the time doing only light work. By the end of this period, you want to be ready for bed.

Ten p.m. to two a.m. is predominated by pitta energy again. The body is certainly on fire, but in a very different way than it was during the day. Now the brain wants to generate ever deeper sleep cycles in an effort to rest and cleanse itself. While daytime pitta energy focuses on digestion, the nighttime pitta slows the digestive process. Instead, your liver and your adrenal glands go to work. Your body is turning that raw nutrition into the hormones and enzymes it needs to function the next day. If you go to sleep early in this cycle, you can get that pitta energy working for your body. But many people stay awake working until midnight or later. They say they get a second wind, and feel suddenly more alert after ten thirty. They don't know why, but I know it's that they like to ride that pitta wave. Staying awake late at night doesn't mean you are a natural night owl. It just means that the urge to sleep is a little bit like a train. It pulls in and then leaves the station at a predictable time. Many so-called night owls would cure themselves of their insomnia (as well as many other health problems) if they would just get on the train by ten thirty. If you do this, your body will thank you in countless ways.

Two to six a.m. is another vata period. Here your sleep becomes lighter and dreams become more vivid. The body is preparing for the active day cycle. If you've ever been awakened in these early hours, you know that you wake up instantly, feeling light-headed instead of groggy. It's the time of quick reflexes and quick thinking again, and some people have the kind of insomnia where they wake up during this period with racing thoughts. I'll talk about combating insomnia in a later chapter. But the thing to remember is that it's easier to wake up in the morning before six a.m. than it is after six a.m., when the kapha energy takes over again.

You can see how the body's needs change profoundly throughout the day and night. Your own schedule is probably at odds with what your body requires to function efficiently. And it's not just eating and sleeping that can be done at optimal times. You can schedule your workload to take advantage of the body's natural energy. I had one patient who told me that she had been in a staff meeting every day at eleven a.m., the pitta time of day. People were cranky and critical of one another every day, a situation exacerbated by the meeting's proximity to lunch. When she learned about the Ayurvedic schedule, my patient moved this daily meeting to two p.m. She told me later that everyone was so calm and creative. It was the same staff discussing the same issues, but at a different time of day they were better able to come up with solutions without arguing.

From an Ayurvedic perspective (and a physiological one), Rory had a couple of major problems. First, he was working out at noon, pushing the blood flow away from his digestive tract and out into his muscles and extremities. You want to think of the noon hour as your biggest opportunity to get dense nutrients into your system. This is when your digestive tract is—or should be—on fire. Replacing that noon meal with a little protein shake is going to cheat your body of all of the nutrients it needs for the rest of the day and night and leave you depleted tomorrow. Better to work out earlier in the day and eat a substantial lunch at noon. His second problem was eating dinner at eight p.m. This is after dark, when the body's digestive system should be powering down. When you overload it with food, that food won't be digested efficiently and will probably sit in your stomach, causing congestion and keeping you awake. His third problem was working late at the computer until midnight. From Rory's perspective, this was an ideal time to work because he had great energy late at night, and few distractions. But your brain needs rest and the best time to settle into that rest cycle is long before midnight. When you look at your own schedule, you want to be thinking of these segments of the day so you can best balance your workflow with your body's needs.

The Benefits of a Healthy Schedule

Most of my patients are used to putting work at the heart of their daily schedule. They keep trying to perfect the workday so that they can get more done. Few people think about how to arrange their day so that their bodies' needs come first. We think of health habits in isolation. You want to lose weight, so you change your diet. Or you want to get fit, so you change your exercise routine. I'm asking you to think of putting all of your health habits into one schedule and make that the center of your day, because these habits all work together. Your sleep affects your weight and your fitness level. Your diet affects your sleep and your mental clarity. Your daily exercise improves your sleep, your energy level, and your food choices.

Still, there are two major benefits for arranging your schedule around the circadian rhythm. Most of my patients want to lose weight. And even if they don't think they need to lose weight, when we change their schedule, they do anyway. The second problem many people have is with general fatigue. It can present as malaise, or low energy, or even mild depression. I think of generalized fatigue as the other national epidemic. But working with your body's clock and correcting problems with the circadian rhythm will also let you tap into new reserves of energy.

Using the Body's Clock to Lose Weight

Adam came to me because, like so many people, he wanted to lose some weight. He told me that he'd gained about thirty pounds over the course of the past ten years, while he was building his construction business. His days were naturally busy, with meetings and phone calls and he drove around for much of every day to different construction sites. Still, the weight gain had troubled him, and he couldn't figure out what was causing it. He was active all day, often doing physical labor, and had cut down on his calories. Nothing was working for him. His wife, who was a patient of mine, had sent him to see me, and he had reluctantly agreed.

When I asked Adam about his day, he told me a familiar story: He had added up the calories for all the food he eats in a typical day, and he often skipped breakfast. He even skipped lunch on some busy days and made up the difference with a midafternoon sandwich from a vendor's truck or a bag of chips.

What was most revealing for me was to hear Adam talk about his favorite time of day, which was the late evening. This is true of many hardworking people. He told me that he worked hard all day, supervising people, checking on details, and sometimes pitching in to get a job done on time. He would get home at night exhausted, sometimes as late as eight thirty, take a shower, check in with his family briefly, and then wait for them to go to bed. This was his favorite time of day, because he had the house to himself. He would heat up a frozen dinner at about ten and stay up watching television until twelve thirty or one in the morning, only going up to bed when he began to feel drowsy or bored. As a result, Adam was getting about five hours of sleep a night, but he would sleep later whenever he could, thinking that he was catching up on sleep on weekends and days off.

He thought he was doing everything he could to lose weight, aside from the corn chips and pretzels. But he needed to stop thinking solely in terms of how many calories he consumed and burned during the day. This wasn't helping him at all.

When you want to lose weight, you have to work with your body's clock by taking in energy when your body can best process it and avoid food when your body is powering down for the day.

Stop counting calories. So many of my patients count calories informally all day long. Most people do. If someone offers you a candy bar, you look on the package to see how many calories it contains. If you want to go out to dinner, you might skip lunch, thinking to save those calories for later.

We even think of exercise in terms of calories burned. Many people have memorized the number of calories you can burn by running or walking a mile, or doing a yoga class. If you get on a treadmill and punch in your weight, the treadmill will provide a real-time estimate of the number of calories burned during your workout. Popular fitness trackers offer the same information about calories burned. Many smartphone apps also help people track calories, grams of fat, and grams of carbohydrates.

However, by focusing on calories, you are treating the body like a simple machine and forgetting that its cells and systems are interacting in complex ways. Focus instead on the timing of your meals and the nutritional content of your food. I'll talk more about this in chapter seven, where I'll teach you how to crave the right kinds of foods for your body.

Become an early eater. Most people following the standard American diet eat their largest meal of the day at night, sometimes late at night, like Adam. It's an easy mistake to make if you think that the body processes all calories the same regardless of when you eat. Obesity researchers are beginning to say that meal timing is the missing link in maintaining a healthy weight.

One recent, dramatic study demonstrated how this might work. Researchers tracked the eating habits of 420 overweight men and women in Spain, who were going to try to lose weight. They were divided into two groups, early eaters and late eaters. In Spain, the main meal of the day is the midday meal, when people consume about 40 percent of their daily food intake. Early eaters were defined as those who ate their largest meal before three p.m., while late eaters consumed their largest meal after three p.m. At first, both groups lost a little bit of weight, but as the study wore on, those eating their largest meal after three p.m. stopped losing weight, despite their healthy diets. Those who ate their largest meal of the day closer to noon continued to lose weight and in the end had lost 22 percent more weight than the late eaters. This was true despite the fact that both groups were eating about the same number of calories overall.[1] This was the first large-scale study to suggest that meal timing affects weight loss. Eating later in the day means that your body experiences a large spike in blood glucose at a time when it is more likely to be stored as fat. But if you eat your largest meal at midday, your body has more time to burn this energy rather than store it as fat.

Give up late-night TV. You may know that brain activity changes dramatically during the sleep cycle, alternating between deep restorative sleep and lighter REM sleep (or dreaming). What you may not know is that the deepest and most restorative sleep cycles occur between ten p.m. and two a.m. This is when the brain is cleansing itself, when the cells of different systems are repairing themselves, and when memory and learning consolidate. Being awake during this time interferes with these important tasks and will leave you foggy the next day. It can also contribute to weight gain. Lots of studies show the connection between sleep loss and weight gain, but one in particular showed that curtailing sleep by even a couple of hours caused participants to increase their daily food intake by more than five hundred calories after just a few days in the experiment.[2] A late bedtime interferes with hormones such as leptin and ghrelin, which signal hunger and satiety in the body, and will lead to overeating.

Exercise before breakfast. Can early morning exercise make up for a less-than-stellar diet? Yes, it can. One study asked men to increase their daily intake of fat by 50 percent and their daily intake of calories by 30 percent for six weeks. One-third of them did this with no additional exercise. Another group was asked to work out in the midmorning, after breakfast. The final group was asked to work out early in the morning, before they had eaten anything. At the end of the six-week study, the nonexercisers had gained an average of six pounds on the high-fat diet. The midmorning exercisers had gained an average of three pounds, while the early morning exercisers had gained almost no weight. Their glucose tolerance levels remained strong and they maintained healthy insulin levels despite their high-fat diets.[3] Exercising early in the day primes the body to receive energy and to process that energy efficiently, whereas intense exercise later in the day offers far less benefit to your metabolism.

Adam needed to eat dinner much earlier and move his largest meal to the middle of the day if he wanted to have any hope of losing weight. He also needed to go to bed earlier. At first he balked. No doctor or nutritionist had ever asked him what time he went to bed or when he ate dinner. While it was true that his blood glucose levels and blood pressure were inching up, he had never gotten any advice about how his daily habits—other than caloric intake and basic food choices—would affect these variables. But by scheduling a longer block of nightly sleep, he could jump-start his weight loss. By eating earlier in the day, he could eat when his body would better digest his food. I also asked him to do a short workout in the morning before breakfast, a brisk walk or some calisthenics to give his metabolism a boost first thing. It's true that he needed to make better food choices, but this is far easier to do after you've seen some results, after you start feeling more alive.

He started losing weight right away, and by the end of three months, he had completely erased the weight gain that a decade of bad habits had given him. Adam's wife reported that he'd stopped snoring at night, and Adam told me that he didn't have any more morning congestion. He felt less bloated even by the end of the first week. He also had a new kind of energy, and he felt more connected to his family. Although he attributed this to the weight loss, I know that one of the major benefits of following the body's clock is an abundance of energy and focus.

Using the Body's Clock to Increase Energy

People often find that when they shortchange themselves on sleep, they are less productive in their work, which makes them feel even more busy. When they feel overwhelmed, they are more likely to stay up late again. Getting your sleep routine down is the first step toward getting more energy.

Martha came to see me because she was suffering from insomnia and a general malaise. An artist who took a day job at a local nonprofit, some nights she was awake half the night doing her creative work, while other nights she was trying to get to sleep early so that she could get up on time for work the next day. To compensate for her lack of energy, she snacked on sugary, carb-heavy treats, particularly on those nights when she stayed up late. She didn't realize that her sleep and eating schedules were so deeply intertwined with her level of fatigue.

By paying attention to your schedule, you can make simple changes that will drastically improve your focus at work and give you more energy during the day. You won't need coffee or sweets to push you through your day if you:

Set a consistent bedtime. Staying up late on some nights but not others sets you up for insomnia. Your body is trying to figure out what time of day it is based on when you get up and when you normally go to sleep. If you don't have a set routine for bedtime, your body doesn't know when to release hormones to make you feel tired and it doesn't know when to start the rest and rejuvenation period in your cells. The result of this is low energy, blunted moods, and poor diet choices. Martha needed to cut off the internet at night and pick a couple of relaxing activities to do every evening.

Exercise every day. It's easy to think about exercise as a means to get fit or trim a few pounds, but it is also a key means of insomnia prevention. The body uses your behavior to help decide what time of day it is. By moving your body in the first half of the day, you reinforce the body's internal cues that it's daytime. Exercise also increases energy if you do it every day. One study looked at sedentary people who suffered from consistent fatigue and found that consistent, low-intensity workouts reduced their feelings of fatigue by more than 60 percent and made them feel more energized throughout the day.[4]

Give up late-night snacks. You might not think that snacking can rob you of energy, but it can. Choosing nutritionally dense foods at mealtimes helps curb snacking between meals. Martha needed to move away from breads and pastas that gave her a short-term rush but ultimately made her more tired. She needed a plant-based diet that would feed her body's systems.

Remember that it's what you eat in the first half of the day that fuels your body for the next day. Everything you eat after dark is more likely to be a drain on your system.

Martha liked the idea of having a set bedtime, even though it made her a little anxious about when she would do her creative work. She hated the idea of joining a gym but found some exercises she could do in the morning. Just a twenty-minute workout was plenty for her body type. She wasn't looking to become ultra fit and therefore didn't need to exhaust her muscles or finish her workout drenched in sweat. She just wanted the jolt of energy that would help wake her up and give her some clarity for the day. Once her sleep cycle settled into a natural rhythm, she found it easier to set priorities at work and she had more ability to focus on her art in the early evening. With a new diet and sleep routine in place, she got excited about looking for a new job and setting new long-term goals.

What Do You Do All Day?

Now, it's your turn. If you want to create a better schedule, you first have to keep track of what you do during a typical day.

Sleep

1. When do you wake up naturally?
2. What time do you turn off your computer and put your phone away?
3. When do you naturally fall asleep?
4. Do you have a different sleep schedule on weekends?

Diet

1. When do you eat your first meal of the day?
2. When do you eat your largest meal?
3. How many calories do you take in after six p.m.?
4. At what time do you eat your final meal of the day?

Exercise

1. How many times a week do you exercise?
2. At what time of day do you typically exercise?

Mindfulness

1. How do you feel before and after you eat each meal?
2. How quickly do you reach for your smartphone when you feel bored or stressed?
3. Is there a time of day when you sit quietly and check in with your body?
4. How often do you move your bowels?

If You Do Nothing Else

Timing is everything when it comes to healthy digestion, restful sleep, and good fitness. The good news is, there's not a lot of guesswork: there *is* an optimal schedule for health. In the coming chapters, you will learn strategies to tap into the wellness potential of your own body type. However, if you only do three things, they should be:

1. **Get to sleep at a set time every night, ideally by ten thirty p.m.** When you do this, you start to feel the effects of greater focus during

the day, often within the first few days. You have a better handle on daily stress. You will also start to lose weight.

2. **Eat your largest meal of the day at noon.** People who eat a hearty lunch have a much easier time maintaining their weight, and many of the digestive issues they have including acid reflux, upset stomach, and constipation go away when they eat at the right time. Most people are used to enjoying their largest meal at night, but this wreaks havoc with the digestive tract. The evening meal should be about half of what you are used to eating. Don't worry about being too hungry. The larger noon meal will fuel you through the afternoon and make those late afternoon snacks and cups of coffee unnecessary.

3. **Exercise first thing in the morning.** Most people don't have to do as much exercise as they think. Spending an hour on the treadmill late in the day isn't going to do as much for you as twenty to thirty minutes of activity as soon as you wake up. Early morning exercise affects your sleep cycle, your weight, and your blood pressure. It discharges stress also. You can get even more out of your morning exercise if you do at least some of it outside, where your brain can bathe in natural light and strengthen your body's natural rhythms.

Do just these three things for seven consecutive days and your health will be transformed.*

---

* After a month, you won't believe you lived any other way.

In the following chapters, I'll explain how to fine-tune these habits to give you the perfect eating, sleeping, and exercise schedule for your body, because every body is slightly different. I'll also offer some tips for getting to sleep on time and setting up the right exercise and diet for you, in case some of these habits seem impossible.

3 Listening to Your Body

Ayurvedic medicine—and the rest of this book—is predicated on your ability to tune in to your body in order to align your body's clock with its natural circadian rhythm. But if you don't know how your body feels throughout the day, it's hard to figure out what or when to eat and when to stop eating. If you are disconnected from your body, it's tough to know when you need rest or exercise. All day, we let our temptations and anxieties lead us around. We are looking to the future and the past for confirmation about why our lives are the way they are. Think of your five senses as five wild horses, while your mind is the hapless driver of a chariot, trying to hold the reins of these horses. Only by tuning in to your body with mindfulness training can you control these wild horses and steer your body and your life in the direction you want to go.

We perceive these changes to diet, sleep, and exercise patterns as changes in the body, but those changes happen first in the mind. When I meet with patients and give them guidelines for resetting their body's clock, they often tell me all the reasons why they can't make these changes. They tell me that they can't give up late-night television, even if it will allow them to have a good night's sleep. They don't want to give up the big evening meal, even if this change will allow them to lose weight for the first time in years. They have a million reasons why they can't exercise in the morning, or why they can't eat on a schedule. You need to find ways to skip over what the mind is telling you about your momentary desires in order to listen to what your body really needs. The biggest battle you will ever fight over your health is not with your body; it is with your mind.

Take Jason, for example. He came to see me because he was having issues with heartburn. The burning sensation in his stomach and throat had become so uncomfortable that he couldn't eat any of his favorite foods. He couldn't eat anything spicy; even the occasional beer would make him miserable. He was just thirty years old and said that he'd never had any problems with food issues, no allergies or anything. After talking a little bit about his diet, he began to tell me about his work. After he graduated from college he had decided to open a business, and he eventually opened a CrossFit studio. Athletics had been a passion for him starting in middle school, and he thought that owning a gym and interacting with people like him all day would make him happy. But his day started at six in the morning and didn't finish until around ten p.m. Like a lot of fitness professionals, he led what seemed like a perfectly healthy life in which the stress was killing him. Jason couldn't sleep at night because of stress, and he couldn't find a relationship. He couldn't remember the last time he went away for a vacation. His business was like a beast that had to be tended to every minute, yet for all that busyness, he didn't feel productive. Like so many busy people, he wanted to change his life but didn't even know how to start. He had no strategy for changing course.

It's not uncommon to have someone come to me with physical complaints that are also emotional troubles. Your body is a complex instrument that reflects back to you who you are. When you become unhappy, your body talks to you about your unhappiness. The more you ignore your body, the louder it talks. Jason's body was beginning to talk very loudly to him. It's possible that it had been trying to communicate important things to him about his lifestyle for quite a while. Unfortunately, he wasn't listening. Only when he found that he couldn't eat from the daily discomfort of heartburn did he decide to seek help.

Although I gave him my usual advice about going to sleep on time, eating at the right time, and toning down his exercise routine, what Jason really needed to think about was the toll that stress was taking on his life. Jason had a lot of passion and perfectionism, but he had lost his purpose, and his body was letting him know this. What he really needed was to check in with himself routinely. He didn't make time to think about how he felt moment to moment, let alone consider whether the life he had built for himself was working for him anymore. Once he'd started checking in with his body throughout the day, he started considering alternate careers. He started asking himself, *Why am I doing this?* And, *How do I feel right now?* And, *What's my purpose?* Even entertaining these questions opened up a new wave of creativity for him. He saw that his fear of failure and of being less than perfect had led him to this demanding work schedule and these health problems. By connecting his mind and body, he began to think in terms of setting new goals for himself. For

Jason, mindfulness, not sleep or diet or workouts, was the first step to transformation.

The False Notion of Willpower

People say all the time that they don't have the willpower to change their diets or to adopt healthier habits. They tell themselves that they should be able to impose a discipline on their minds and bodies by force. But this effort is exhausting and so it fails. Throughout the day, you build up stressors or toxins in the mind. If you don't have a way to discharge them, then you will reach for distractions of every kind. It doesn't matter what your temptations are. They could be internet games, buying cookies you know you shouldn't eat, reaching for an extra beer, working too much, obsessing about things out of your control. When your stress level builds and you don't discharge it, your temptations will be calling you. You will feel the need to mask your discomfort with unhealthy habits. Those wild horses will be dragging you every which way, and you can't beat them into submission. We call this *bhoga*, or indulgence in luxuries and comfort. An excess of bhoga leads to *roga*, which means disease.

You can break the cycle of bhoga and avoid roga by pulling the senses under control. And you do this with *yoga*. While you may think that yoga is a series of exercises, the word really means union. It is the union of body and mind that breaks the spell of the senses and allows you to have stable thoughts and a stable intellect. Spend a few minutes every day checking in with your body and you will see temptations and unhealthy habits fade away. You won't have to force them to go. They will just lose their power. I've seen this countless times with my patients.

The very first step in doing this is taking your own pulse.

Listen to Your Heart

As an organ, the heart symbolizes so many things. When we talk about our hearts, we may be referring to feelings of tenderness, passion, gratitude, sympathy, or grief. These emotions may be generated in our minds, but we feel them in our bodies. That's why a connection to the body is so important. In Ayurveda, taking someone's pulse is a way to find out how they really are, which energies are dominant, or how many toxins may be lurking. We also use it to assess emotional well-being.

Taking your own pulse is fairly easy. You can find the pulse point on your wrist. Men should use the left hand and take the pulse in the right wrist. Women should use the right hand to take the pulse on the left wrist.* With your palm facing up, wrap the opposite hand around the back of your wrist so that the first three fingers are on your pulse point near the heel of your hand. Make sure you grip the wrist from behind, so that the forefinger is closest to the heel of your thumb. You can do this right now if you like. Close your eyes if that makes it easier to focus on the feeling of the pulse and sit for about twenty or thirty

---

* In Ayurveda, we associate the left side of the body with lunar energy and the right side with solar energy. Women's bodies are more in sync with the lunar cycle, in that their hormones fluctuate every twenty-eight days, which is why they take their pulse on the left wrist.

seconds. You aren't trying to time the beats or measure them in any way. Just notice them. Feel your heartbeat and the breath in your body rising and falling. It's very peaceful. Sometimes you will note that the pulse feels stronger in one or another of your fingers. If you do this several times a day, you will discern the natural changes in your pulse at predictable times. It will feel different before and after you eat, before and after you exercise, and it will certainly be different just before bed or when you first wake up in the morning. Over time, you will sometimes be able to feel your heart beating without taking your pulse, and you may get used to becoming still enough to feel all sorts of sensations in the body. For example, you might feel the mild contractions moving through your digestive tract.

You can also attend to the way that stress manifests in the body through aches, dullness, a fog, or a sharp pang. You will learn by focusing on what kinds of things are contributing to the body's sense of stress, and also those things that bring joy. Above all, you will learn to appreciate your body and how it is working all the time on your behalf. You may realize how much you love your body and how much you want to take good care of it. All sorts of insights can come from this one exercise, which is a kind of meditation.

At first, you will be checking in with your body at predictable times. It's a way of seeing how you feel throughout the day, which will give you valuable information about how your behavior affects your body. For example, if you are struggling to lose weight, you will want to take your pulse before and after each meal. Again, you aren't looking for a particular rate or change. Your job is to notice the heartbeat and to notice your body, to notice how you feel. Asking yourself how you feel before you eat is another way of saying, *Am I hungry?* And afterward, *How do I feel after eating what I just ate?* These are revealing questions, because people sometimes say to me that they can't give up bread or even cut down on bread because they love it so much. And then I ask them to check in after eating it. And what they find (especially those with some insulin sensitivity) is that the bread still tastes fantastic, but within twenty minutes their hearts are racing and they feel anxiety, or a kind of restless panic. And this is not unusual. If your body has trouble processing simple sugars, this is its normal reaction to that. Once they experience this feeling, they are better able to cut down on things that are making them sick, even if they taste good in the moment. Eventually, your taste buds will change and those things won't even taste good anymore. The cravings will disappear. Until then, you want to be checking in with your body. It's talking to you all day long, but you haven't been listening.

If you have insomnia, you want to check in as you go to sleep at night and when you wake up in the morning. You will also want to check in during the day whenever you feel overwhelmed, because feelings of anxiety during the day can reemerge when you are trying to sleep. If you are starting a new exercise regimen, you want to check in before and after exercise. This is how you know if you are doing enough exercise, and particularly if you are doing too much.

Many of my patients use this simple tool as the starting point for mindfulness. It seems like a funny little chore at first, but within a few days they rely on it more and more. It's amazing how quickly you can dissolve stress by closing your eyes, taking a deep breath, and feeling the steady pulse of your heart.

Stress and Screens

How many times do you reach for your phone during the day? In any public place, you can see people staring at their screens, not because they've just received an urgent call or text. Quite the opposite. All day long we turn to our phones for something to do, and it reliably gives us a shot of emotion, frustration, anxiety, or good feelings when you get a text or email from someone you wanted to hear from. We tie ourselves to our phones. Or we look at social media feeds and click on things to feel a certain way, to feel angry about the unfairness of the world, or to feel good while looking at cute animals playing. That's fine. But I want you to spend some time disconnecting from all of that. Remember that the desire to reach for your phone is your mind urging you to seek a shot of adrenaline or dopamine to cover up for its momentary discomfort. Instead, take a deep breath and resist. Ask yourself how you really feel. If you ask your body, *How are you feeling?* your body will tell you.

Interoceptive Awareness

If you can balance your thoughts and listen to your body, you are in the right frame of mind to change your life. Checking in with your body and attuning yourself to its state is a part of every healing tradition. And now Western science is catching up with this ancient wisdom. Researchers don't always call it mindfulness training, or meditation. Instead, they focus on what they call interoceptive awareness, or awareness of the body's physiological function.

It's not easy to measure how well people perceive their own

physiological state. You can't just ask them how well they are tuned in to bodily functions. Instead, you have to give people a task that involves bodily awareness. In one study, researchers decided to divide subjects into two groups. The first had to listen to a series of notes and decide which one had a different pitch than the others. The second group had to say whether the series of notes sounded in sync with their own heartbeat. The subjects also had their brains scanned with a functional MRI during the exercise. The act of focusing on their own heartbeats caused one particular area of the brain to light up the scan.[1] This is the *insula*, a small part of the brain that helps translate bodily sensations into emotions. And these emotions drive many of our behaviors. The insula has only been studied for about a decade, but it seems that this same area of the brain serves a different function in humans than other mammals. In mice, for example, the insula turns bodily sensations into instinctual behaviors. In humans, it turns them into desires, cravings, and habits. Of course, it can also turn sensations into emotions, such as passion, disgust, and fear. This is the part of the brain that turns physiological sensations into subjective emotions.

Some people have a strong ability to tap into this physiological awareness, and most people can learn to do it, but what does this do for us? Some studies showed that this kind of practice helps people cope with anxiety.[2] It helps addicts in recovery to withstand cravings.[3] It helps people with insomnia to sleep at night, and to help them relax even when they can't.[4] What I've noticed is that it helps people to better interpret and reflect upon the reality of their lives. It gives them a little speed bump, let's say, between the urge to do something and the action itself. It is the most powerful lifestyle change you can make, and very few people do it.

People assume that they can use willpower alone to change, that they can use their minds to force their bodies out of bad habits. Actually, your habits work in reverse. Your body is giving you all kinds of clues about what it really needs, but over many years you have taken those cues and turned them into subjective emotions in the form of cravings and habits. In order to have a chance at breaking those, you have to tune in to the body and its source of raw data. Taking your pulse several times a day is an easy source for this data. Over time, you will want to set aside more time in your day to do more formal meditation techniques.

The Power of Mindfulness

One of my patients is a senior manager at work who came to me because he was struggling with stress, weight gain, and some eczema. Although he followed the exercise and diet recommendations, he resisted the idea of meditation, saying that it seemed counterproductive and wasteful. He also thought it had to be done in the morning, which was hard to work into his schedule—he liked to be at the office before everyone arrived and stay at work into the evening. When he realized that he could meditate at the end of his workday, he agreed to give it another try. He would sit down on the couch in his office and do a short meditation before his commute home. His family noticed the change in him right away. His wife said he looked different when he walked in the door at night and his kids noticed that he wasn't irritable at the dinner table anymore. He felt more present for his family and less consumed by work and, once it became a permanent part of his routine, he told me that those twenty minutes were the most important change he'd ever made.

This story is not unusual. I've seen people who told me they would never give up smoking and drinking. They believed that they couldn't do it because they had tried and failed so many times. When they try simple meditation techniques to become more connected to their bodies and the way their thoughts swirl around and often misinterpret the body's cues, they find that they can slow down and feel centered more often throughout the day. Then when I check in on them months later, they tell me that they have given up all their bad habits and completely transformed their diets. Sometimes these lifestyle changes cause them to feel differently about themselves, which leads to new behaviors at work, new ambitions, and better relationships at home. Corporations have also included mindfulness training with great results. Google has a mindfulness course with a six-month waiting list, because graduates rave about its powerful transformation of their lives and careers.

In order to appreciate mindfulness training, you first have to understand the body's fight-or-flight response. When you become upset or anxious, your body reacts as though you are facing a physical threat and need to run away or fight an opposing force. In a sudden emergency, you can probably feel the adrenaline hit your system. Your heart rate increases along with your blood pressure. You breathe more quickly and your mouth becomes dry. You may perspire slightly in acute stress. Your muscles become tense as though ready to pounce. There are also things happening inside your body that you can't feel. Your digestion slows. Your body releases stress hormones, such as cortisol, which causes inflammation. Your blood sugar levels increase while your immune response slows. The body prepares itself to handle emotional stress as though you are facing a physical threat.

In an ongoing emotional crisis, your fight-or-flight response may be active for a large part of every day. Over time, though, your body can learn to respond to the world as a kind of threat and keep that stress alive long past the point where it is useful. In fact, stress can become an emotional habit. You may tell yourself that you feel stressed out, but you aren't really feeling your body at all. Instead, you rush from one activity or deadline to the next. Many people live in a near-constant state of low-level alarm in which they have slightly elevated levels of blood sugar, blood pressure, and inflammation. Their bodies are tense, their jaws hurt from gritting their teeth, and their backs and necks ache from this constant tension. There is some research that suggests that the part of the hypothalamus that controls your circadian rhythm can be disrupted by stress.[5]

Mindfulness training is a way to reset the fight-or-flight response. It allows your body to experience its natural state of restful awareness. By focusing on your breath and listening to your body, you are pausing the body's stress response. Your pulse and blood pressure can go down while your tension melts. Adrenaline and cortisol levels subside and your blood sugar decreases. You may also find that you sleep better at night and eat better during the day. You can get all of these benefits from sitting and doing nothing.

The best part is, you don't need to sit on a cushion or chant, or listen to audio tapes, unless those things appeal to you. There are many ways to achieve mindfulness and, depending on your preferences, sitting meditation, movement meditation, or journaling—or a combination of the three—can prove equally beneficial.

Sitting Meditation

This is the most traditional form of mindfulness and one that I have taught for many years. The steps are simple, and doing this for even a few minutes every day can be very beneficial.

1. **Choose your mantra.** Choose a word or phrase to silently repeat to yourself during meditation. This will give you something to put your attention on other than your thoughts. You can use "om" as a mantra, which means being present and connected to the field of consciousness. You can also use words, such as "peace" or "joy." Avoid words that are too loaded with meaning for you. The word "love" is sometimes a challenging mantra for people, because it means so many things. Remember that a mantra is just a tool to help focus and quiet the mind. Some people don't use one at all. They choose to count their breaths instead.
2. **Sit comfortably.** Find a quiet location where you won't be disturbed. There is no need to sit cross-legged on the floor. You can sit on a chair or sofa or on the floor with your back against a wall. You may support yourself with cushions, pillows, or blankets. The goal is to sit as upright as possible while remaining comfortable. We all have different anatomies and you want your meditation experience to be enjoyable, so make your comfort a priority. Remember that meditation can be practiced anywhere.

3. **Gently close your eyes and breathe.** Take some deep cleansing breaths. Inhale slowly through your nose and exhale through your mouth a few times. Then continue to breathe normally with your mouth closed.
4. **Repeat your mantra.** You can do this silently without moving your tongue or lips. Repeat the word slowly. There is no need to rush or to try and time the word with your breath. Allow the breath to fall away into its own rhythm. The repetition of your mantra should be almost effortless. The mantra will help you transcend to restful awareness. As you continue, you will find that you drift away from the mantra. The mind will wander and there is no need to empty your mind or try and stop your thoughts. When you notice that you have become distracted, simply return to repeating the mantra.
5. **Let the mantra go.** After approximately fifteen to twenty minutes (you can use a timer with a very gentle sound to let you know when to stop), you can stop repeating the mantra and continue sitting with your eyes closed for a few moments before you get up.

You'll notice that your state of mind after you finish meditation will be very different from the way you normally engage with the world. Your thoughts may be slower and you may feel less pressured and harried. With practice, you will be able to remember this state of mind intermittently throughout the day. You don't need to try and re-create that sensation all day. Just know that it's available to you whenever you sit down to meditate.

Movement Meditation

When people tell me that they can't do a formal sitting meditation, I ask them to try using movement instead. Some people have trouble sitting still because they think they aren't accomplishing anything. Or they can't yet feel a change in their state of mind while sitting and breathing. You will feel these things with practice, but there are other types of mindfulness techniques you can use in the meantime. We use movement meditation when teaching kids about mindfulness, because they are often so full of energy that any effort to sit down and breathe makes them giggly and wiggly. Instead, we give them a word or phrase to repeat to themselves silently and have them walk around in a circle. Within a couple of minutes they are quiet and centered, and this feeling lasts for several hours.

Walking meditation. Some people have a strong urge to walk when they feel anxious or need to think, but this is called pacing. Meditating and pacing out of anxiety aren't the same thing, but you may be one of those people who does better while walking. With a walking meditation, you take a walk outside, perhaps in nature, and instead of ruminating on all that you have to get done or all of the conversations you would like to replay, you just walk and breathe. Notice the weather and the nature around you while you breathe deeply. In Japan, walking in nature is sometimes called "forest bathing," and I love that term because it captures the powerful sense of peace you can attain by walking among trees. The Japanese government has spent millions of dollars on research documenting the positive effects of forest bathing on the immune system, and its ability to reduce stress, inflammation, and even blood pressure.

While doing a walking meditation, you don't need to worry about getting exercise, so you should move as slowly as you like. You are looking for a shift in focus from your swirling thoughts to a complete connection with your breathing and the movement of your body. As with seated meditation, the goal isn't to suppress your thoughts, but rather to notice them and let them go. Your job is to focus on your body and your breathing. You will, of course, get distracted by insistent thoughts, but you have the power to turn your attention back to the steps you take and to your breath and surroundings.

People who struggle with anxiety and perfectionism often find that this kind of break in the day takes the edge off. Others, who are stuck in the past, either with grudges or regrets, find the ability to let go for a few minutes at a time. This can be extremely therapeutic. When I lived in Hawaii, I had a friend who was a psychiatrist. Although he was semiretired, he still saw some patients. But he would meet his patients on the beach. He told me that walking along the beach felt restful for both doctor and patient and it allowed people to more easily unburden themselves. He would walk with each patient for about ninety minutes and let them talk. Their thoughts ranged widely from problems to their ambitions. And sometimes they would be silent for stretches of the walk. He was an unusual psychiatrist in that he preferred to practice outdoors, and he also gave his patients books to read and talked to them about spirituality as well as psychology. But it was the walking that helped them the most. They were in a beautiful setting, talking to someone who was actively listening to them, and they were allowed to be silent and have their own insights. He said he never needed more than four months to get them back on track.

Gentle movements. If you've ever stood up during an intense meeting or work session just to stretch your arms or roll your neck, you've done this type of mindfulness work already. The most formal version of this is tai chi, which encourages you to move and breathe using specific forms. It is a wonderful way to gently build the discipline of self-awareness. Any gentle movements or light stretches can work also if you tune in and pay attention to how you feel.

Many people use yoga as a form of meditation. These gentle but formal stretches ask you to coordinate movement with breath and to hold positions that may be unusual for you while feeling all of the sensations in your body. For people who struggle to meditate, these movements and poses allow you to focus completely on movement and holding a position. It is the ultimate distraction from modern life. Holding a pose is like holding your breath. It may produce a little bit of anxiety, but it focuses your attention entirely on what you are doing instead of your thoughts about the future and the past. I often recommend yoga classes to people who want to combine mindfulness with exercise.

This sometimes causes resistance. I had a patient recently who needed to lose some weight and start an exercise routine for the first time in many years. I suggested a gentle yoga class and she told me that she didn't even like looking at her body in the bathroom mirror. How could she possibly put on shorts and a small T-shirt and do a bunch of barefoot exercises in front of strangers? I had to tell her that it's just a class. The worst thing that can happen is you show up, breathe deeply, and try to move your body in new ways. Yoga stretches can put you in touch with small and large muscle groups you haven't thought of in years. It flushes your system and brings profound relaxation.

And yet I knew what she feared. Getting in touch with the body you have neglected is a complicated business. It's easy to feel competitive with others or to feel left out if you can't do all of the movements the way the instructor does. Some yoga studios do cultivate a sensibility that is intimidating. Perhaps they attract very competitive students who strive to overachieve by taking back-to-back hot yoga classes. Some yoga students do confuse physical flexibility with spiritual enlightenment. Or they start thinking that they need a designer outfit and high-tech mat in order to look the part. Or they attach themselves to a particular instructor and treat this person like a guru. All of these things run contrary to the purpose of yoga, which is to bring your awareness to your body. If you look around, you can find the right class for you.

Journaling as Listening

Writing down your thoughts is a way of capturing them, but it also serves as a way to let them go. You can journal at a specific time of day, or you can carry a journal with you and jot down notes a little at a time as a way of checking in on yourself. You can ask questions about whether you are happy, what daydreams you have. You can think about goals, what you are looking for in an ideal job or partner. In a way, a journal functions as a kind of trusted friend who will listen without judging or offering advice. You can ask yourself any question: Why can't I follow this diet? Why don't I exercise more? Why is my relationship not working? What kind of job would I like? Any answer you come up with is a starting point in this dialogue with yourself.

Journaling is a great way to begin to answer important questions about what you want out of life. Frequently, I meet with people in their twenties who feel they have lost their way. Sometimes this has to do with body image. They've had a poor diet or exercise routine, or they have trouble sleeping. Often it's because they don't know what to do next in life or how they managed to end up in careers they hate and they wonder how to change anything at all about the future. I sometimes ask them, "Pretend I have a magic wand that I can wave and give you anything you want. What is it?" And they look confused. They say, "I don't know." This is the problem. It's your job to figure out what you want. It's mother nature's job to manifest things for you. I don't have a magic wand, but the journal is the closest thing I can give you. Here is where you can begin to discover who you are and what you want. It lets you know what you think about all day. It accepts without judgment all of your pressing concerns and your dreams.

The habit of journaling can help you let go of worrisome thoughts and it can help you reset your life goals. That's what Jason used his journal for, when he was struggling with his CrossFit business. Although he took his pulse regularly and did some meditating in the early hours of the day, it was the journal that helped him to focus. He could use downtime at work to jot down ideas about other businesses to start, and about the kind of life he would like to lead. Gradually, he began to fill more and more pages with his philosophy about training, motivation, and personal discipline. Ultimately, he was surprised to find that he had the beginnings of a book on the subject. This became his next venture, the place where he could find purpose in his life.

Fasting for Mindfulness

In Ayurveda, as in many healing traditions and religious traditions, fasting is a means for enhancing a spiritual practice. I'm not suggesting going days without food, or even a whole day without food, although for many yogis, this is a way of attaining spiritual awareness. However, going without food for a few hours is good for your body. We do this naturally every evening, "breaking our fast" in the morning. During the day you should also think of the times between meals as a time to abstain from food. We live at a time when we can eat all day and all night if we like. But going without for three to six hours is a good way to check in with your body by leaving it alone. You practice this by eating breakfast by eight thirty and abstaining from food until twelve thirty. Then after lunch, you eat nothing until dinner at six thirty. Going without for a few hours gives you the chance to ask if you really need a particular food before you eat it.

Some people deepen this practice by choosing one day a week to skip their evening meal. The evening meal is a good choice, because your body doesn't need much fuel at this time of day. By giving up one meal a week, you can experience the lightness that comes from eating less. If you do this, make sure to drink warm water with lemon or herbal teas in order to keep the body hydrated throughout the evening.

Seeing Toxic Buildup

Tuning into your body is more than just quieting the mind. It also means noticing how your body is functioning on a physical level. In particular, you want to be aware of the toxin buildup in your body. In Ayurveda, these toxins, called *ama*, are the byproduct of unhealthy habits and emotional trauma. Ama accumulates more quickly when you are out of tune with your body's circadian rhythm. When these toxins build up in the system, they can cause weight gain, inflammation, aches and pains, or other diseases. If you have trouble losing weight or chronic aches and pains, you know you have built up ama in your system and your new healthy schedule will help release it.

Your body is naturally inclined to release ama when you move your bowels, when you sweat through exercise. But when your schedule overrides these natural opportunities, ama will continue building up in your body. There are simple ways to notice the amount of toxic buildup in your system, and taking the time to look for toxins will give you a good measure of your progress in changing your schedule:

Look at your tongue. Go to the bathroom mirror every day and stick out your tongue. Look at the surface of your tongue. You might notice a white or yellowish coating. In some cases, the coating might have a greenish tinge. Any coating indicates the presence of toxins that your body is trying to flush from your system. This would also indicate that you still have some changes to make in your daily diet. You are still eating some things to which your body is reacting badly. And bacteria is accumulating in your mouth and on your tongue as a reaction to these things. It could be that you are eating too many simple sugars, too much oily or heavy food, and not enough clean fruits and vegetables. It could be that you are still giving in to cravings for junk food and unhealthy snacks.

One way to help your tongue shed these toxins is to use a tongue scraper. Silver and plastic varieties are available online, though I recommend a silver scraper: while both do a mechanical job at loosening the toxins, silver also has antibacterial properties. Just draw the scraper over your tongue, not hard enough to scratch or hurt the surface, but with enough force to gently draw out the fluids trapped in there. When you do this, you are drawing out the bacteria and toxins on the surface of the tongue and spitting them out. Do this every day, first thing in the morning, and over the course of a couple of weeks, you will see the coating gradually diminish. Many dentists now recommend brushing your tongue daily to cut down on the bacteria that contributes to the buildup of plaque and irritates the gums. Scraping your tongue offers these same benefits, but it also serves as a reset button for your taste buds. By eliminating the bacteria and the microscopic residue of your old diet, you are actually helping to retrain your taste buds. Over time, cravings for "bad" foods will diminish and it will be easier to change your diet.

Regardless of whether you get a scraper, the real point is to look at your tongue every day. Get in the habit of seeing the results of your daily dietary choices. Many people look to the number on the scale or the waistline as the ultimate judge of whether they are following a diet. But your body is much more than your weight or your waistline. To be truly connected to your body—and to make the most of your body's natural cycles—you need to know how it responds to the food you put into it. And the first place to see the effects of your diet is in your mouth.

Look at your stool. Take a peek into the toilet every morning after you move your bowels. This sounds unpleasant. It may seem even more unpleasant to have me tell you that your stool should be one elongated piece, shaped like a banana, and that they should come out of you easily in one go. They should also float. You may protest and say that never in your life have you produced such a thing. You may produce hardened little clumps every few days, which means that you are probably dehydrated and not getting enough good oils in your diet or enough fiber. You may sometimes produce a loose mess that smells awful, in which case you might be eating things that cause an allergic reaction in your bowels, or you may be developing difficulties digesting lactose or simple sugars. The same is true if you are producing more gas than stool. This is your body talking to you about what it can and can't digest well, and you should listen. Every day your bowels should be moving because every day you are putting more food into your system. If nothing is coming out the other end, it's because your body can't process yesterday's food, and whatever you ate yesterday is still fermenting in your system. The result is a buildup of ama in your system that makes it even harder for your body to absorb the good nutrients you do put in. Obviously, this is not good for you.

When people change their diets and try to eat more cleanly, their bowel movements change immediately, but they don't always become smooth, perfect specimens. You may have an abundance of toxins built up in your system, and these will need to come out in their own time or with a little help from the detox diet outlined in chapter seven. So you may have some mornings with sludge or looseness until your new diet takes hold. Eating a lighter meal in the evening, fasting during the night, and getting a good night's sleep will also improve your bowel movements and cause them to arrive first thing in the morning, right on schedule.

People sometimes tell me that I don't understand how their bodies work. They have never moved their bowels regularly no matter what diet they have tried. So I sometimes challenge patients to take a picture of their bowel movements every day. This is how they keep a record of the changes in their bodies, the reward for all of their hard work to change their lives. I admit that this kind of record-keeping is not for everyone. Even so, I encourage you to try it, because you have verifiable proof that your body is changing along with your new regimen. If a photo is too graphic for you, perhaps you can keep a note in your journal alongside your food diary and exercise log. Your bowel movements are just as important as either of those things. Keep a record of how you do in the bathroom, and I promise you surprises as you change your lifestyle.

In fact, patients often send me emails, the subject line of which is: I DID IT!, usually accompanied by a detailed description of the bowel movement they just had, its consistency, the fact that it's floating. They sometimes send a photo as proof of the wondrous bowel movement, which I encourage them to do. At this point, my daughter is afraid to borrow my cell phone or even to touch it, because it contains so many close-ups of patients' tongues and bowel movements. People send me Picasso albums of their thirty-day log of tongue coatings as proof that they are checking in with themselves every day as I've asked them to do. They say that they never thought they could do it. But what they are really saying is that they never thought they would care so much about how their body functions day to day. But now they do.

Resolve today to put some of these techniques into practice and you will see how much easier it is to make changes to your sleep, diet, and exercise routine. Mark on your calendar, or set an alarm in your phone to remind you to take your pulse or do a short meditation every day. Even five minutes of meditation will feel restful and rejuvenating. It will reintroduce you to your body. Mindfulness is the golden key to change because it connects you to the signals your body is giving you about what it needs. Every experience you have creates a buffer between your body and mind. Meditation clears that buffer and takes you back to the yoga, the union of body and mind.

4 Sleep Is the Miracle Drug

How are you sleeping at night? This may be the most telling question my patients answer during a first consultation. From an Ayurvedic perspective, sleep is as important to good health as the food you eat and the air you breathe because the body cannot thrive while deprived of rest. If you don't balance your activity with rest, you will deplete your strength, weaken your digestive fire, and ultimately shorten your life span. Not only does sleep provide rest and rejuvenation, but it separates you from the illusions provided by the five senses. In sleep, you are transported to a distant field of awareness where your ego is dissolved and you exist in a purer state. As you sleep, your stressors are released into that field while your slumbering body repairs itself and awaits your return. In Ayurveda, sleep is considered a spiritual experience, one that you shouldn't curtail in order to watch a little TV or send a few extra emails.

Even so, many people seem to have resigned themselves to a life without good sleep. Patients tell me that they wish they had more time to sleep, or that they lie awake at night wishing they were sleeping, or that they rely on sleep aids to get enough hours of sleep. They have turned sleep into a mystery, and they hope that they are getting enough to function at their peak during workdays.

Researchers in Europe, who have been tracking sleep patterns of working adults, have found a disturbing trend: for the last decade, people have gotten four minutes less sleep each year. Four minutes may not seem like much, but it adds up. If you do the math, on average you are probably sleeping about forty minutes less per work night than you did ten years ago.[1] People are staying up later or lying awake longer and yet they still have to get up at the same early hour to get to work on time. My patients who struggle with sleep tell me that the biggest problem with waking up sleepy is the feeling that they can't concentrate for the first couple of hours at work. Also, they find that they aren't hungry in the morning, but by noon they are ravenous and snacking on everything. They crave junk food and sweets and caffeine. Their weight creeps up, they think, because they can't stick to a diet.

In reality, it's the lack of sleep itself that changes your metabolism and makes weight loss impossible.[2] First, it reduces your resting metabolic rate, which is the amount of energy your body uses throughout the day just to function. This may be because your body senses that it needs to conserve energy when it hasn't had enough rest. So, when you are operating on a sleep deficit, you are going to be burning fewer calories in a twenty-four-hour period, regardless of your level of activity. At the same time that your body is burning less energy, it's going to be craving those starchy, sweet snacks that can deliver a jolt of energy to the system but contain no nutrients. Again, this may be linked to the body's sense that it needs to conserve or hoard energy. What's more worrying is that a lack of sleep can disrupt your body's ability to process energy, and particularly to process those simple sugars you start craving. You can experience significant insulin resistance after just five days of reduced sleep.[3] This means that by the end of a hectic workweek with lots of late nights, your body is primed to store fat. Even if you get extra sleep for the next three days, which is longer than most weekends will allow, you probably won't be able to restore your insulin sensitivity to normal levels in that time. By contrast, getting good sleep normalizes your metabolism and reduces your cravings for junk food. If you've ever fantasized about losing weight by doing absolutely nothing, I've got good news for you: pick an early bedtime. It's the best diet you'll ever find.

Sleep is also a powerful anti-inflammatory. People who experience a chronic sleep deficit have larger inflammatory markers in their bodies. So if you suffer from chronic pain or are at risk for any cardiac issues, you probably need more sleep. I tell my patients that if you aren't getting to sleep at the right time, no diet will make you thin and no exercise regimen will make you fit. You will be more susceptible to colds and flu during the winter months. Over time, you are more likely to develop metabolic disorders and heart disease.

Tracking Your Sleep

People often say to me that they need more sleep or want more sleep, but when I ask them how much sleep they get on average, strangely enough, they don't know. Perhaps you do know that the alarm goes off too early on workdays, and you know that you tend to stay up later on weekends because you can sleep in. You may also know that you stay up too late some nights and struggle to fall asleep after you've turned out the light. But it's important to note that sleep is a habit, not just for your mind but for your body. Looking at the different sleep patterns you have for workdays and days off can give you important clues about why you have those Monday-morning blues and why sleeping late on weekends actually creates insomnia during the week.

Sleep researchers are newly interested in the timing of sleep and how it varies between workdays and weekends and across geographic populations. Researchers in Germany came up with a questionnaire to assess people's sleep habits and to measure how they differ between workdays and days off. It's called the Munich ChronoType Questionnaire (MCTQ)*.[4] It's an interesting premise, the notion that people have different natural sleep rhythms and that these can be measured and quantified by a questionnaire. The MCTQ has been used by researchers the world over to gather information about sleep habits. Hundreds of thousands of people have taken this questionnaire, filling out their age, weight, height, and sleep patterns. With this information, researchers can glean insight into the correlation between sleep schedules and obesity, for example. That's also how we know that bedtimes for working people have been progressively delayed during the past decade, even though work start times haven't changed. People aren't necessarily working until late at night. Yet they still stay up late and get up early, and many of

---

* You can take the quiz yourself at: https://www.thewep.org/documentations/mctq/item/english-mctq-core.

them sleep in on the weekends or nap to make up for lost sleep. You can't change your sleep patterns until you know what they are, so the first step is to start tracking your sleep cycle and ask yourself some of these questions:

1. **On days when you know you have to get up and go to work, what time do you go to bed? And at what time do you get up?** These questions can help you figure out how much sleep you get on a work night, and they can be revealing. Many people are up until midnight, even on workdays when they know they have to get up in a few hours. They know the alarm is going to go off. And yet they tell me that they can't fall asleep before the "usual" time. This is the direct cause of your sleep deprivation on workdays. It interferes with your mental functioning the following morning, so you aren't getting as much done. It changes the way you eat on workdays. Many people who feel groggy in the morning don't eat breakfast and then wonder why they eat so much at lunch or snack so much throughout the day.
2. **On work mornings, how much do you rely on an alarm to wake up?** About 80 percent of respondents to the MCTQ say they need that alarm to go off in order to get up.[5] This is another sign of sleep deprivation, but it also shows that the body is out of alignment to its natural rhythm.
3. **On work mornings, how long does it take you to get out of bed after the alarm goes off?** For some people, the alarm goes off and they are wide awake. This may be a sign that you are a light sleeper, or it may mean that you have a good sleep regimen already. Others need that

snooze button. And they are dragging even after they get out of bed.

You may be telling yourself that you are a night owl and that this morning drag is normal. But there is something else at work here. The body has its own natural sleeping pill, called melatonin, a hormone that the pineal gland releases in the brain to make you feel sleepy. Melatonin circulates throughout the body and helps communicate to organs and tissues that the sleep cycle has begun. It begins to release in the evening and peaks about two hours after you fall asleep, at which point, it decreases. About an hour or so after it leaves your system, you wake up naturally. If you are groggy in the morning, it means that your body's melatonin production is out of alignment with your work schedule. There is still a lot of melatonin in your system and trying to get up is like trying to function under the effects of a sleeping pill. If you can't fall asleep at night, you may have delayed melatonin production. If you can't get up in the morning, you may still be under the effects of latent melatonin. Don't worry. There are things you can do to shift this, and we'll get to those in the next section.

4. **On free days, what time do you go to bed? And at what time do you wake up in the morning?** About 70 percent of respondents to the MCTQ say that they alter their sleep patterns by at least an hour on the weekends. They go to sleep an hour or so later at night and sleep an hour or so later in the morning. Further, about 30 percent of respondents say that their sleep patterns change by two hours or more on the weekends. They stay up much later at night because they know they don't have to get up

the next day, and in the morning, they sleep in an extra two hours or more. Perhaps you do this, too, thinking that sleeping in will leave you refreshed to start another workweek of sleep deprivation. It sounds logical. Unfortunately, it doesn't work. In fact, altering your sleep patterns during the weekends creates a natural pattern of insomnia during the first few days of the workweek, and this leads to sleep deprivation. It also contributes to weight gain and digestive issues, and makes you more prone to stress.

Social Jet Lag

If you sleep in on the weekends and get up early during the week, then your social sleep schedule and your work schedule are out of sync. Researchers call this phenomenon "social jet lag" because it mimics the effects on the body of traveling across time zones. By staying up until midnight all weekend and trying to reset in time for work at nine a.m. Monday morning, you are essentially flying one thousand miles west on Friday night and flying home again on Sunday afternoon. Experienced travelers know what this does to the body. Jet lag causes sleep deprivation and mental fogginess, sure. It also causes digestive issues. Travelers sometimes suffer from an upset stomach, constipation, or just feeling bloated and unwell. People who constantly change time zones also get run-down more easily, are more susceptible to colds and flu, and are more sensitive to emotional stressors. These travelers know that when they come home, their bodies will settle down again. If you are living with social jet lag every weekend, your body never gets a chance to normalize.

The concept of social jet lag and its effects on the body is a new area in research, but studies have already shown that it contributes to metabolic disorders. People who have a body mass index that is above normal are at higher risk for developing obesity and type 2 diabetes if they experience social jet lag over several years. They also tend to reach for substances, such as caffeine and alcohol, to moderate their wakefulness and deal with stress.

Researchers also tracked the ages, heights, and weights of people who filled out the questionnaire and they found that sleep deprivation is most acute during adolescence and young adult years and declines fairly steadily until retirement age. And yet, social calendars and work calendars are most responsible for sleep interruption. They also note that although people can sleep outside their circadian window for sleep (nighttime), other researchers have found that these extra sleep times don't provide good quality sleep. Naps and sleeping in on the weekends, for example, don't provide the quality of rest that you can get by sleeping when your body expects to sleep. Instead, it's a heavy sleep that leaves you groggy. Also, the day's circadian rhythm as set by the SCN starts when you wake up and see daylight. If that's at nine or ten in the morning, your body will track that as your wake-up time for the next morning. You won't begin to feel sleepy for another twelve to fifteen hours. Your digestion, body temperature, blood pressure, hormone levels, and cortisol levels are all still trying to operate on that natural twenty-four-hour cycle. They are up and moving on a weekend morning even if you are not. They are powering down after sunset on a weekend night, even if you are dancing at a party. You may feel up and awake, but many of your body's systems are desperately trying to get some rest. Over the course of a couple of days, your body can get used to a new set of times, and it will reset your body clock and all of your systems accordingly. Scientists call this entrainment. But staying up late for two days and then going back to your usual schedule does much more than make Monday morning a drag. It sets off a cascade of effects in your body. It changes the way your body functions. Most immediately, it inhibits your weight-loss goals.

The first goal in changing your life is setting an appropriate bedtime and sticking to it throughout the week. I always suggest you get to sleep by ten thirty. This is the end of the kapha time, according to the Ayurvedic schedule, which means your body is naturally a little bit heavy and sleepy. After ten thirty, though, you are in the pitta time of day, when you are likely to get hungry and awake. I always say that ten thirty is the last bus leaving the station to get to sleep.

What Happens When You Sleep?

Despite all of the research on the benefits of sleep and the problems associated with not getting enough, scientists still don't know exactly what's going on when we sleep. Or why we do it. We know a lot about it, but sleep is still a little bit mysterious. We know that during the first two hours of sleep, the cells in your brain are working to discharge cellular debris. Throughout the rest of your body, you use the lymph system to flush out debris, but the brain has no lymph system. It needs sleep to carry out this process.

We also know that sleep comes on gradually in stages. The earliest stages are light. About ninety minutes after you fall asleep, you enter your first stage of rapid eye movement, which is also called REM sleep. This first REM cycle lasts about ten minutes, but each subsequent cycle lasts longer and the final REM stage may last up to an hour. During REM sleep, your brain is more active and you may experience dreams. Between these cycles, you enter non-REM sleep, which is when your body repairs its cells and strengthens its immune system.

You need at least three of these REM and non-REM cycles per night, but four or five is better. When you are staying up late, you are getting a late start on this important brain activity, and you are probably hitting your deepest and most restorative levels of sleep just as the alarm is going off at six a.m. You'll know that you are doing this if your alarm goes off and you wake up feeling groggy and disoriented. You may think it's because you are not a morning person, but it's really because you've cheated your brain and body out of the restorative sleep it needs. Everything chronobiologists are learning about sleep supports the Ayurvedic view that getting to bed at the right time is the best way to take advantage of sleep's essential benefits—even those that are still mysterious to us.

Getting the Melatonin Flowing

I have lots of patients who tell me that they would love to go to bed earlier, but they don't feel sleepy until after midnight. They tell me that they are "natural" night owls. I sympathize with that feeling, but for most people this simply isn't true. While it is true that you may not feel drowsy until late at night, that's thanks to years of training your body to delay your natural sleep cycle. You have tricked your body into releasing melatonin later in the evening, and that delay prevents you from getting drowsy at the right time. It's a kind of invisible insomnia and it has real health consequences.

Carla came to me complaining of weight gain and insomnia, thinking that the two problems were unrelated. She was particularly puzzled by the extra pounds, because she was often eating salads at work and frequently logged seven miles or more running around on her job as a manager of a large Costco store.

A closer look at her schedule revealed a problem that many working adults have. She was delaying her entire evening routine. If you are having trouble feeling sleepy at ten or ten thirty, you may have created a schedule that precludes drowsiness at the right time. Many people start this syndrome by working late at the office, even when they don't have to. Instead, they tell themselves, *I'm just taking care of a few extra things.* Then dinner is delayed until eight or eight thirty p.m. Then the restlessness sets in. They scroll through channels on the TV, check on social media, do some internet shopping. Or they answer meaningless emails in the name of doing more work.

All of these activities—working late, eating late, engaging in electronics late at night—have the same effect on the body. They delay the onset of natural drowsiness. They delay the production of melatonin in the body. That's a problem, because melatonin isn't just a substance that makes you feel sleepy. It also circulates in your blood. It communicates with the organs and systems in your body and with your clock genes to signal that the rest and restoration period of the day has begun. Your body is really busy during your sleep cycle. It is clearing debris from cells, producing hormones and enzymes that you need to function optimally the next day. You need that melatonin to build up in the first couple of hours after sunset, and it needs to reach a critical point long before the evening news begins. This is how your body synchronizes its clock genes with the brain's central clock.

If you are wide awake at eleven or midnight, you have created a disruption in your circadian rhythm whether you know it or not. Most people, like Carla, see its effects first in stubborn weight gain, but it contributes to everything from heartburn to congestion to digestive issues and cardiac problems. It also causes a near-constant state of fatigue. Many people live with this desynchronization for years.

For Carla, setting an evening bedtime meant giving up evening TV and finding something to do with those extra two hours after her eight thirty TV curfew. Eventually, she settled on light housework, journaling, a nice long bath, some reading before lights-out at ten thirty, and then a short meditation exercise in bed. Within a few weeks, evenings became her favorite time of day. Instead of searching for mindless TV to distract her, she was focusing on herself and her goals. She found renewed energy for socializing and hobbies in the early evening and on weekends. Of course, she also lost weight. In fact, she lost more than the fifteen pounds that had originally brought her to my office.

Many of my patients think they have a weight problem or a diet problem, but when I put them on a good sleep regimen, they lose weight. Sometimes they lose ten pounds, but others have lost up to thirty pounds. Everything is easier to do when you feel refreshed in the morning. It's easier to make good food choices and to go to the gym. It's easier to deal with stress at work and at home. When your body works in sync with the cycle of light and darkness, all of the body's cells and systems are performing as they are intended to. This creates an explosion of new energy.

Natural Light: The Missing Ingredient in Sleep

Most of us live indoor lives. In winter, we can have days in which we wake up in darkness, sit in a cubicle far from natural light, and then travel home from work in the dark as well. This takes an emotional and physical toll on our bodies.

Natural light is the primary mechanism by which the body sets its daily circadian rhythm, and when we don't get enough natural light, the body gets confused about when to be awake and when to sleep. When someone says to me that they aren't sleeping, and that they are feeling sluggish in the morning, I say, "Put on your shoes and take a ten- to twenty-minute walk every morning." If you do this, you will be the most wide-awake person at the office. And you will sleep better every night.

As far as what you can do at work, I advise taking work breaks outside. Walk around the block instead of having a cup of coffee. Or take your coffee with you. Even if you can work near a window for a couple of hours during the day, the quality of your sleep will improve. Some of my patients use special alarm clocks that are also lamps. Instead of using sound to wake you up, the lamp slowly lights up and floods the room with full-spectrum light. They like it and say that it helps them to get up in the morning. Others use full-spectrum lights at their desks as a way of getting more natural light during the workday. I think these are useful devices, but nothing really replaces the act of getting outside or near a window for a couple of hours a day. Natural light improves your sleep, but it also improves your mood and outlook.

This was recently shown by a small study out of Northwestern University. Researchers tracked the sleep habits of people who worked in offices with a lot of natural light and compared their responses to those who worked in cubicles and had low exposure to natural light during the day. Those who worked in an office with abundant windows and natural light received 173 percent more white light exposure during work hours.[6] They also slept for an average of forty-six minutes longer per night than those without. People who worked in places with windows were more likely to get regular exercise, while those who worked in offices without windows reported poorer quality of sleep and more sleep disturbances. Overall, these workers had less energy and more physical complaints.

Even if you can get outside for just a few minutes at a time, a couple of times a day, you will find your outlook improving along with your quality of sleep. I've heard from people who say their health and sleep times improve after they adopt a dog. And this makes sense to me, because the one thing you have to do every day is walk the dog. You have to get out and walk at least around the block first thing in the morning. You may have to walk the dog several times a day, which means you are getting more time outside in natural light, which is going to help you feel sleepy in the evenings and reset your circadian rhythms. I wouldn't suggest adopting a dog if you don't have one and don't want one, but rather that you remember that your body has needs, too, and be sure to take *yourself* for a walk.

Night Lights

This idea of shutting off the TV and hiding the clicker after eight thirty or nine p.m. is a tough one for many people. But sleep isn't something that just happens to you. It's something that you and your body have to prepare for. When I tell people that they need to power down their electronic devices in order to lose weight and clear up their health issues, they balk. One man was nearly in tears when he said to me, "Don't take my late-night shows away. It's all I have." On one hand, I do understand that modern life is stressful. I've heard from so many people that their favorite time of day is late in the evening, when the kids are asleep and when no one can reasonably expect you to answer work-related emails. The house is finally quiet, and you just want to turn on the TV and zone out, or turn on your e-reader and sink into a book. Or perhaps you want to play video games or do some internet shopping. It's what they call "me time." But the ultimate "me time" is what you give to your body so that it doesn't break down.

Nearly every American adult uses some type of electronics at least a few nights per week, and by this I mean cell phones, e-readers, or computers. (And those who don't like to text and send emails at night are probably watching TV to unwind.) But what do these electronic devices actually do to the brain? They offer just a fraction of the amount of natural light you would experience by taking a walk during the day. How bad could it be? Studies show that using electronic devices at night has a negative effect on sleep. One theory is that these devices have low wavelength-enriched light, meaning that the light emitted by these devices contains more blue light than natural light. It turns out that people are highly sensitive to this blue light, which makes the body more alert. Even though lights from electronic devices aren't as bright as natural light, their saturation in the blue spectrum means that they work on the brain in a similar way as daylight. So, if you are sending emails or playing games on your phone late at night, you are suppressing melatonin production and shifting the body's clock.

In one study, twelve adults were given either e-readers or traditional books and told to use them during the four hours before bedtime. After four days of this, the subjects using e-readers reported that they were less able to fall asleep at their normal bedtime. It took them on average ten minutes longer to fall asleep compared to those who read a traditional book in dim light. This doesn't seem like much, but in the morning, those same participants using e-readers felt less awake and said that they needed more time to feel awake in the morning.[7] Additionally, researchers found that they showed less REM sleep, even though they had slept the same number of hours as the people who read from paper in reflected light. Their blood work showed that they had lower melatonin levels at bedtime than those who read traditional books, suggesting that your body suppresses melatonin production while you use an e-reader or other electronic device the same way it does during daylight hours.

It may seem odd to think about using an e-reader for four hours straight, but if you think about all the electronic devices that you use over the course of an evening, including cell phones, tablets, and computers, it's not as far-fetched. Some cell phones are far brighter than e-readers, and people tend to hold them closer to their eyes than they do with an electronic book. So cell phones can be even more disruptive to your sleep cycle. Many teenagers and young adults move from tablet to e-reader to computer to cell phone all day and night. I wonder how much trouble they will have sleeping as they move into their adult years if they don't know how to power down and prepare for sleep.

No matter what, you need to power down about two hours before you expect to fall asleep. Your brain needs time away from the artificial light emitted by electronics in order to start the flow of melatonin. And you need a break from stressful work or media that is playing on your emotions. This is a time to become quiet in body and mind so that sleep can come on gradually. People often say to me that this sounds lovely, but then they ask: What can I do instead?

1. **Crack open a book or magazine.** Now is the time to read the novels on your nightstand or to work through the back issues of your magazines. One patient of mine gathered together all of the industry journals he was supposed to read and put them by his bedside at night, using the time to catch up on work reading he had been struggling to find time for. After the first week, his insomnia disappeared.
2. **Try oil massage.** You can use sesame oil or regular olive oil to give yourself a light massage. Even if you just want to massage your feet, this can be extremely relaxing. The oil itself seeps into the skin with its linoleic acids, and in addition to moisturizing the skin, it acts as a natural relaxant.
3. **Take a bath.** Many of my patients have turned to evening showers because the warm water helps them shed the stress of the day. For those people who struggle with insomnia or have trouble calming their minds during a stressful time, a bath can be even better. Some of my patients take a bath after they give themselves an oil massage and find that this is the time of day they look forward to the most. They go to bed feeling clean and relaxed.

4. **Try some warm almond milk.** A few people say that they feel hungry at bedtime and this makes them restless and unable to wind down. After the first week with your new bedtime, this feeling of hunger should subside, but until then, you can heat up a cup of almond milk (or regular milk) and sprinkle in some saffron or a little bit of cinnamon. There's something about drinking something warm and creamy in the evening that helps you relax.
5. **Write in a journal.** One of the reasons Carla couldn't sleep was that she became anxious in the evenings. That was one of the reasons she had for turning to television. It was a reliable distraction from unwanted feelings. When you feel separated from your life goals, it's always tempting to turn to distractions. Instead of running away from your feelings, you can write them down. Some studies have shown that writing about your problems or about a distressing event for just twenty minutes helps you discharge your negative feelings and feel more peaceful, even if the problem isn't yet solved.[8] Now Carla writes regularly and says that the habit helps her to relax and to think creatively about her goals and how to handle tricky situations at work.

Let There Be Light

In sleep labs all over the world, scientists are studying the effects of sleep disruption on the circadian rhythm and the body. Very few researchers are studying ways to get a great night's sleep. And that's too bad, because that's what we are all looking for, isn't it? Thankfully, there are some researchers who are interested in the kinds of conditions that can help boost the body's circadian rhythm. In one recent experiment, researchers were able to dramatically improve the sleeping patterns of study subjects, just by taking them camping. In this study, a group of people were sent on a camping trip in the Rocky Mountains of Colorado in July. They had been observed for a week of their normal routine in which scientists tracked their sleep and wake cycles. Researchers also used little devices called Actiwatches to track their daily activity and observe how much light they were normally exposed to. For the six-day camping trip, these folks were asked to leave their cell phones at home, along with all flashlights and lanterns. They were encouraged to use campfires as a source of light after dusk, but they could have no electric lights. When they returned, the researchers found that the campers had dramatically improved their sleep cycles.[9] They went to sleep more than an hour earlier than usual and woke up closer to dawn without any alarm. Also, the melatonin production in their brains had changed. In the days after the camping trip ended, participants released melatonin about two hours earlier than usual. It peaked earlier, during the first half of their sleep cycle, and then waned a couple of hours before dawn. What's significant is that before they went camping, participants were going to sleep at about twelve thirty a.m. and waking up at about eight in the morning. After the camping trip, they tended to fall asleep by ten thirty p.m. and wake naturally at first light. Even more interesting was the fact that the campers experienced four times the amount of light while camping than they did in their normal summertime routines, despite the lack of electric devices in the wilderness.

This might sound obvious. Of course, they went to sleep earlier! After all, on a camping trip you start a fire at sunset, cook your food, and then sit around looking at the stars while enjoying light conversation. There's no television drama to turn to, no emails to check, and no work deadlines to anticipate.

It's unrealistic to think we can solve insomnia with a week off the grid in the mountains; however, it is fascinating how quickly the body adapts to the natural light-dark cycle when given the chance. The melatonin levels in your brain—the body's strongest natural signal for sleep—will adapt and work with you if given the chance. And this change can happen faster than you think.

Encouraged by the results of this first camping trip, researchers decided to try again. This time some participants were sent to the same mountain area near the winter solstice, when days are shortest, while another group stayed at home to be monitored living their regular routine. The experiment lasted one weekend.[10] After just two days, the winter campers had transformed their sleep cycles and brain chemistry. They went to sleep earlier, enjoyed longer and deeper sleep, and woke at dawn without an alarm. Despite the short winter days, the campers were exposed to about thirteen times more natural light than their counterparts who stayed at home. Think about that. The people who stayed at home and had electric lights in every room and closet, outdoor lighting, television, flashlights, and cell phones still experienced a fraction of the amount of UV light that people enjoyed while camping during the shortest days of the year. The campers' brains released melatonin earlier, and the amount of melatonin peaked earlier and then diminished in the predawn hours, making waking up that much easier. By contrast, the people living their normal routines might have melatonin levels starting to rise at about ten thirty at night, and that means they wouldn't begin to feel sleepy until after midnight. The melatonin might peak between two and three in the morning and would still be affecting them when they woke at eight a.m. It's a peculiarity of modern life that many people feel groggiest an hour or two after they wake up. Superimpose this melatonin production on a workday, in which you stay awake past midnight because you don't feel sleepy and then the alarm goes off at six a.m., when the melatonin production in your brain is still winding down.

The most important thing this study shows is that the brain can right itself very quickly. Within forty-eight hours, the campers had completely reset their circadian rhythm. For many people, insomnia is a result of too much fake light in the evening and not enough natural light during the day. In fact, the absence of natural light may be a major factor in why people can't get to sleep on time. So, if you are still having trouble getting to sleep on time, try getting more natural light during the day, particularly in the morning.

Mealtimes and Sleep

Mealtimes can have a huge effect on your ability to sleep. Eating a heavy meal later in the evening or snacking at night will interfere with your body's ability to relax. That may seem counterintuitive, because eating a big meal can make you feel groggy. In reality, your body can't digest food at night. The digestive process slows to a crawl after dark, so any food you take in will sit in your intestines and ferment. This causes gas, stomach pains, and heartburn, which leaves you tossing and turning. What's worse, your body may produce mucus or congestion as a result of eating late in the evening. Physicians who specialize in treating acid reflux often tell patients to eat their evening meal earlier in the day—at six p.m. instead of nine p.m.—in order to reduce acid reflux. They also know that related symptoms, such as congestion, persistent cough, sinus trouble, and even allergy symptoms, get worse when people eat later. (Heavy desserts, including ice cream, are particularly disruptive to sleep.) These symptoms resolve remarkably when you eat your last meal of the day in time for your body to digest it.

Eating a light dinner at six p.m. will do wonders for your sleep. Giving up late-night snacks can also boost the quality of your sleep by reducing stomach issues at night.

Although having the right sleep routine has many benefits for everyone, good sleep has particular benefits for different types of people. You may have specific sleep challenges, depending on your body type. In the next chapter, I'll help you figure out what kind of sleeper you are and give you some hints for working with your body type to get the best restorative sleep for you.

5 The Right Sleep Routine for You

The quality of your day depends on the quality of sleep you got the night before. That means you have to know how to prepare to get a good night's sleep, and that preparation depends on what kind of sleeper you are. Most people think of themselves as night owls or morning people. But it's a little more complicated than that. You might be a light sleeper who has trouble falling asleep or a heavy sleeper who struggles to wake up. Some people fall asleep easily and wake up in the wee hours out of stress or restlessness. There are actually three body types, or *doshas*, in Ayurveda that describe your body's tendencies with regard to sleep, diet, exercise, and outlook. While all of our circadian rhythms depend on the sun, it doesn't mean that everyone's body responds identically. In Ayurveda, knowing your body type is essential, since it informs how you deal with stress, how you transition from sleep to wakefulness, and how you react to various temperatures and foods. So, while it's essential for everyone to have an evening routine to promote sleep, not every routine will look exactly alike. And that's okay. You want to work with your body and not against it. Once you have your sleep routine in place, you will notice an increase in your level of energy and focus. Your mood will improve. You may also start to lose some weight.

The quiz below will help you identify your sleep habits and can tell you a lot about how to get a better night's sleep. Remember, there are no wrong answers to these questions, and it's okay if there are a few questions in which none of the answers fit exactly. Try to find the answer that comes closest to describing you.

1. **My natural frame would be considered:**
   a. Thin or small.
   b. Strong or tall.
   c. Stocky, regardless of height.

2. **When I complain about the temperature of the room, I'm usually complaining that:**
   a. I'm cold, or my hands and feet are cold.
   b. It's too hot in here. I'm sweating.
   c. It's stuffy, or too humid.

3. **When I have insomnia, it's usually because:**
   a. I spend so much time trying to relax enough to fall asleep.
   b. I wake up hours early and can't get back to sleep.
   c. I feel physical discomfort at bedtime or during the night.

4. **In physical terms, I know I will have trouble sleeping if:**
   a. I'm cold. Even under the covers, I can't get warm.
   b. The room is too warm or I wake up sweating.
   c. I can't find a comfortable position.

5. In emotional terms, I know I will have trouble sleeping if:

a. My thoughts are racing with excitement or dread, or I'm replaying a conversation that didn't go well.
b. I have a pressing deadline or feel overwhelmed by work. I have too much to do.
c. I'm worried about my health or someone close to me who is struggling.

6. If I try a new diet, I know I will wake up at night because I'm painfully hungry.

a. No.
b. Yes.
c. Sometimes.

7. In the first hour after waking up, I feel:

a. I'm awake, but still pulling it together. I need coffee or my routine to help me get focused.
b. Wide awake. I'm checking email before my feet hit the floor.
c. Mornings are tough. I like to start slow.

8. On those nights when I stay up later than I know I should, it's usually because:

a. I get a second wind after ten p.m. and get excited to start a project or movie.
b. Even though I'm tired, I'm trying to cross more things off my to-do list, or I can't let go of a pressing project.
c. I'm out with friends and can't break away.

9. **My partner says I keep him/her up because:**
   a. I toss and turn.
   b. I fling off the covers.
   c. I snore.

10. **When physical discomfort keeps me awake, it's usually because:**
    a. I'm achy or my legs are restless and active.
    b. I have heartburn.
    c. I can't get the pillows right so that my neck feels comfortable.

11. **When I have a bad night's sleep, my biggest challenge is that I feel:**
    a. Spacey and exhausted, and not able to function.
    b. Irritable because of low energy. I'm off my game.
    c. Groggy or congested at first, but I can push through it by midmorning.

12. **When I wake in the night, it's often because:**
    a. Any noise, dream, or change in light can wake me.
    b. I have a pressing problem that won't leave me alone.
    c. I don't know what wakes me up.

13. **When I do wake up one to two hours before I have to, I will:**
    a. Try to go back to sleep, because rest is important to me.
    b. Start thinking about everything I have to get done.
    c. Rejoice, because I have more sleeping to do.

14. **My natural disposition could be described as:**
    a. Gregarious, inquisitive, or anxious.

b. Driven, decisive, or active.
c. Easygoing, serene, or generous.

15. Even though it's a little unfair, the consistent complaint I hear from loved ones is:

a. I'm trying to do too many things, or I don't finish what I start.
b. I can be a little bit bossy or defensive.
c. I can't make a decision, or I give in to keep others happy.

16. In terms of the seasons, my mood and sleep patterns change in the winter because:

a. The air is so dry, I get uncomfortable and restless.
b. It's easier to sleep when the house is cooler.
c. It's really hard to get up when the sky is still dark.

17. My mood and sleep patterns change in summer because:

a. I feel happier and sleep better when there is more light.
b. I love the light, but the hot days can make me frustrated and irritable, and unable to sleep.
c. I love summer, but sometimes struggle with seasonal allergies. Humid days are awful.

18. When I dream, I often dream about:

a. Someone chasing me or some kind of disaster.
b. Action-oriented dreams, making plans, or rehashing what happened during the day.
c. Big bodies of water.

TOTAL: A: ____ B: ____ C: ____

If you answered A to most questions, you are probably a light sleeper. If you answered B, you are what I call a variable sleeper, and if you answered mostly C, you are a strong sleeper. It's possible that your answers are mixed. You might answer half in the A column and half in the B column, for example. That's okay, too. Some people have mixed sleeping patterns. It means that you should read both of the following sections that apply to you.

Light Sleepers

If you answered A to most of the questions, you are a classic light sleeper. In Ayurvedic terms, you would be called a vata type. "Vata" means air. Like the air, you are constantly shifting and moving. You have a wide range of interests, make connections between disparate things, and may work in a creative field. Sometimes you may have trouble making up your mind or deciding on a course of action and may be easily distracted. In terms of movement, you may be prone to fidgeting, talking with your hands, or walking quickly, which is a reflection on your quick-moving thoughts. Think of a butterfly, beautiful but with a tendency to flit from one thing to another, always seeking novelty. You may talk a lot and be naturally social. If you are shy, you may have lots of ideas about what to say, even if you don't say much. You love multitasking and don't prioritize neatness, because you have so many things in your life that you love, and because you are often doing ten things at once. In terms of physical characteristics, you are more likely to be naturally thin, and you probably have skin that is prone to dryness. In Ayurvedic terms, we think of these light sleepers as being changeable, like the air around us. Your energy comes and goes in bursts, like your appetite and your mood. To you, the future is one big open country, and you can't wait to get there and so you are always making plans and thinking about contingencies, even if they aren't very practical. You are always asking: What if?

So what does all of this mean in terms of sleep? As you might guess, anyone who is constantly moving in life is going to struggle settling down in one position to sleep. The biggest problem that light sleepers have is with overstimulation. You're likely to be entranced by the internet, dramatic movies, or exciting texts from friends, which disrupt your regular schedule. While you wake up easily in the morning, getting to sleep is a continual challenge, which may make you more likely to find yourself wide awake at midnight, engaging in distractions to put off bedtime. Light sleepers are the ones who most often turn to sleep aids because they can't wind down. Yet these same people struggle the most with the aftereffects of sleep aids, often becoming scattered the morning after taking them. Light sleepers often don't get enough rest during the day, and they exhaust themselves. Ironically, this makes it harder for them to go to sleep early enough to get the rest they need. They may suffer from restless leg syndrome or have aches and pains from overexercising that keep them up.

If you're a light sleeper, here are some things you can do to fall asleep on time:

Minimize stimulation before bedtime. Though all body types need to minimize screen time before bed, it is absolutely essential for light sleepers. But you also want to be careful about all kinds of emotional stimulation. You need to unplug from emotional conflict, violent television shows, and internet shopping sprees for about two hours before bedtime. No work emails or texts. They just get you wound up at the time when you need to be winding down. Instead, you need to think of this as your personal time to recharge. And because you are multitasking and exhausting yourself all day, you need this time more than any other body type. This has to be a firm rule. Any surge of adrenaline will get the mind churning, and it takes time to wind down and relax. You need to be vigilant about turning off the phone, tablet, and e-reader as well. Light sleepers are especially vulnerable to the lights on small screens. This blue light will advance your circadian rhythm in ways you don't want. Instead, you should think in terms of relaxing music, a little oil massage, or some light reading. If you are feeling lingering conflict from your day, discharge it by writing in your journal. This will allow you to make your thoughts concrete, so that you don't have to keep obsessing about them. A journal is also a good place to brainstorm about the future, something light sleepers love to do.

Warm your hands and feet before bed. Light sleepers hate to be cold. Being cold actually triggers anxiety, and yet your hands and feet are often cold because of poor circulation. Warm yourself with an oil massage or just soak your feet in warm water. Some of my patients consider the evening bath or shower an important luxury because it warms them up and allows them to become drowsy. You may also need multiple layers of blankets so you don't wake up cold.

Start an evening meditation. Lots of studies show that a little meditation during the day helps alleviate insomnia. When you can observe your thoughts, you are better able to let them go. Also, regular meditators tend to feel less anxious about the fact that they can't fall asleep immediately. Try to schedule a sitting meditation every evening. You can start with five minutes and work up to twenty minutes a day. For those who don't meditate, I suggest some evening journaling, which does many of the same things. If you really struggle to go to sleep at night, you can try a technique called *yoga nidra*, which is essentially meditating while lying down in bed: Lie down in bed, turn out the lights, and, instead of letting your thoughts run wild, focus on your breathing or count your breaths. Unlike a sitting meditation, you don't have to set a timer and you don't have to worry about falling asleep at the end—that's the goal. Focus on your breaths and notice if there is any tightness in your body while you breathe. After a few minutes, you will drift into a relaxed, slightly sleepy state, which is powerfully restorative. When you practice this over weeks or months, you may experience a lightness or gratitude that comes with meditative states. It can bring insights and creativity and a sense of connectedness to the world.

Keep your bedroom dark and silent. Light sleepers toss and turn throughout the night, looking for that one warm and relaxing position. Your partner might say that sleeping next to you is like sleeping next to a flopping fish. In addition, you wake up with every slight noise, every change of light, every startling dream, and then have trouble settling down again. Your mind is on high alert, even during sleep, and so having a dark and silent environment is critical. You may need to invest in blackout shades to keep your bedroom dark.

Hydrate your body and your house. You may notice that your insomnia becomes worse in the winter months, when the air is so dry. Your skin and nasal passages may become irritated. Your watchword during the winter months is "hydration." Divide your body weight in half and you will know how many ounces of water you should be drinking every day. Avoid caffeine in the afternoon and alcohol at night, because these further dehydrate your body and make you less able to handle stress. You may want to invest in a small humidifier for your home or bedroom. In the winter, your diet needs to contain some oily foods, including nuts, fish, and cheeses. A completely low-fat diet isn't good for light sleepers, because they need these healthy fats to stay hydrated and grounded.

Why You Should Give Up Sleep Aids

A few years ago, a patient came to me complaining of insomnia. Actually, he was concerned about his habit of using prescription sleep aids to get to sleep on some nights. He didn't use the pills every night, and yet he wanted to stop using them altogether, because he didn't like the idea of relying on a pill to get to sleep. He is a successful surgeon, and he didn't feel as sharp as he wanted to the morning after taking a sleeping pill. On the other hand, he had an exceptionally busy and stressful life and he worried as well about getting the right number of hours of sleep. Because he was a doctor, I didn't have to tell him that taking a sleeping pill can make it harder to fall asleep the night after you take the pill.

Depending on who you talk to, sleep aids might be a brilliant invention, or they might be a modern scourge. My patients refer to them as both. They love the convenience of taking a pill and having some confidence that they will feel drowsy in a short time, and yet they hate having to take a pill in order to do something as simple as fall asleep. None of this surprises me. Doctors write about forty million prescriptions for sleep aids every year, and yet half of adults say they can't get adequate sleep. Prescription sleep aids seem so seductive because they deliver the desired drowsiness quickly. This seems like a bonus to people who only want to think about sleep in the few minutes before they go to bed. This is especially true for light sleepers who forget to wind down before bed. However, nearly everyone who uses these medications tells me that they never want to use them again. If you do use sleep aids, it's important to know what effect they have on your brain. You need a better handle on how your body prepares for sleep in order to give them up, since you have specific sleep challenges based on your body type that you need to address to get to sleep on time.

Not all sleep aids are alike. There are over-the-counter pain relievers that double as sleep aids, but there are also two common kinds of prescription sleep aids. It's important to know how they work in order to understand how they don't really support a good, restful night's sleep.

Benzodiazepines in small doses cause muscle relaxation and can reduce the anxiety that causes insomnia for some. In larger doses, they inhibit memory, coordination, and emotion. They also interfere with rapid eye movement (REM) sleep and keep you from getting a deep sleep, which is where tissue growth and repair occur in the body.

Selective GABA medications (Ambien, Lunesta, Sonata) bind with some GABA receptors in the brain to bring on drowsiness.

Both of these classes of drugs are controlled substances, because of their propensity for overuse. My biggest concern about them is that they both have a long half-life in the body. A half-life is the amount of time that your body needs to metabolize half of the dose. Your liver may still be metabolizing a benzodiazepine after ten hours, and a few of them require more than eighteen hours to be out of the body completely. (Zolpidem [Ambien] is the exception, with a half-life of two and a half hours.) If these pills have been combined with alcohol, they stay in your system even longer. Of course, as you age, the body needs even more time to fully rid itself of the drug and its effects. These aftereffects can include grogginess the morning after, difficulty completing complex tasks, and memory impairments, the same side effects that people report after not sleeping at all.

People who are struggling to sleep prefer sleep aids that work quickly, in part because most of my patients think about sleep in the few minutes before they want to be asleep. The truth is that sleep starts when you wake up. You enhance it by making good choices every day to exercise, to eat at the right time, and to prepare for sleep before bedtime.

When you are trying to put away the prescription sleeping pills, you can use melatonin, which is an over-the-counter supplement. It's not a sleeping pill, but something that puts some of the body's natural melatonin into your system at the right time to encourage sleep. Melatonin is a hormone released by the brain that causes you to feel drowsy. Although it's released in the brain, it actually circulates in the blood. So you can take an oral dose that will encourage your body's natural drowsiness. But melatonin also helps coordinate the clocks in your tissues and organs to the master clock in the brain, so that the sleep you do get is more restorative for the whole body. Melatonin doesn't work exactly like a sleeping pill. It doesn't cause you to go to sleep. Instead, it supports your body's natural triggers for drowsiness. And it gives your body a strong signal to shift into the sleep cycle. I tell people to take these supplements at or just before the time they would like to sleep in the next few nights. If you want to become sleepy at ten p.m., then take the melatonin supplement at nine thirty or ten p.m. for two or three nights in a row and you will notice that you become naturally sleepy at that time. (Melatonin has other benefits for your digestive tract, and I'll talk more about that in chapter six.) It's not a solution to insomnia because you need to prepare for sleep by eliminating electronics, eating earlier in the day, and getting more natural light to prepare to get to sleep. Still, it can help train your body and mind to prepare for sleep while you are working to give up sleeping pills.

Variable Sleepers

If you answered B to most of the questions, you are what I call a variable sleeper. In Ayurveda, this is known as a pitta type, and "pitta" means fire. If this is your sleeping style, you may also have a fiery nature. People may say that you are intense, but from your perspective, you just want to get things done right. As a variable sleeper, you may be strong or muscular by build and have a naturally warm body temperature. When you are intensely focused on a task or when you are in a disagreement with someone, your body heats up along with your emotions. You are likely to be task-oriented, someone who makes lists and uses lists all throughout the day to make sure you get things done on time because, if you don't accomplish what you want, you can get frustrated and irritable. This is even true on vacation, where you bring lists of all the things you want to see and experience. You may be a natural leader and a natural public speaker. People listen to you. You can be witty and enterprising and can accomplish almost anything.

Variable sleepers are motivated, persuasive, and committed to their habits. These are the people who live with chronic sleep deprivation and don't understand why their bodies won't just drop off to sleep when ordered to do so. With so much to do and not enough hours in the day, they stay up late to get more done. Even if variable sleepers do fall asleep from sheer exhaustion, they often wake in the early hours of the morning. This is when stress or frustration catches up to them. Many of my patients who can't sleep are up watching TV and hoping that this will relax them enough to trigger drowsiness, but it doesn't. Instead, this commitment to "relaxing" mixes up their internal clock, so variable sleepers often have no desire to sleep before midnight. They often eat late at night, too, because they always need food. Anyone who works as hard as they do burns through a lot of calories, and they may find that they are painfully hungry at bedtime when they try to restrict their food intake for any reason.

Variable sleepers also struggle to sleep in any room that is too warm. They are the ones who need to have their feet outside the covers or throw the covers off during the night because they are too warm. Stress and worry heat up the body, and they have trouble sleeping without a fan or an open window to keep their body temperature stable.

If you're a variable sleeper, you may find it helpful to try some of these strategies as you build your sleep routine:

Step away from work at nine p.m. sharp. Variable sleepers don't like to be told to do less work, but it's the one thing that can transform your sleep schedule and your life. Set an alarm on your smartphone if you need a reminder to unplug on time. Even if you don't use those late-night hours to work, you are probably using television as a means to force yourself to relax, and you think you need TV because you've spent twelve hours marching yourself through endless tasks. You can't unwind with TV late at night. Instead, it's keeping your adrenaline stoked up and pushing sleep away. You need to unplug from all of that and spend some time with your family or with yourself. You've worked hard and achieved so many things. You need time every day to appreciate them. Set a routine that involves puttering around the house, light reading, an evening shower, a gentle hobby, or meditation. Write down a list of things to do if you have to. Many variable sleepers need a to-do list. It isn't easy to stay away from work goals and pressures and mindless media in that ninety-minute period before bed, but it's so important. My variable sleepers fight me so hard on this, but once they make the change, they can't believe they ever lived any other way. When you take away your self-induced sleep deprivation, you will have more time for your most pressing goals and you will function better at work.

Work out before breakfast. Intense individuals need intense workouts. If you have this body type, you love to sweat and you love competition and these things are good for your body. I'll talk more about this in chapter nine. Unfortunately, many of my variable sleepers exercise after work because that's the most convenient time for them. If you are coming home sweaty from the gym at nine p.m., your body is going to stay overheated for a couple of hours and that will keep you awake. You can't sleep when you are too warm. You are also infusing the body with stress hormones, such as cortisol, adrenaline, and norepinephrine, and these can keep you awake. When you work out before breakfast, you are jump-starting your body before your workday begins, so you can hit the ground running first thing at the office. You also release pent-up heat from your body, which will make you more calm during the day. Moreover, variable sleepers have a strong appetite, and that intense, early morning workout will keep your metabolism humming so you can stay trim.

Keep cool. Variable sleepers hate to be too warm. You are the sleeper who needs to leave your feet uncovered at night and who throws off the covers in the middle of the night. You need to cool off before bed. That might mean bathing your feet in cool water or keeping a glass of water by the bed, or you might need a quick, cool shower before bed. You might need to leave the window open or a fan going most nights.

Prevent heartburn. Variable sleepers also struggle with heartburn at night. And they need a diet that will reduce heartburn. That means reducing saturated fats and desserts at dinner, while eating as early as possible. Your body will need several hours to digest dinner, so I suggest eating a light dinner by six p.m. if you suffer from heartburn. For variable sleepers, who go through a lot of calories in a day, this can be a difficult adjustment at first. I tell these folks that they can eat half of a banana, or a small piece of fruit later in the evening if hunger pangs become too intense.

Don't fret about waking up early. Variable sleepers do wake up early, especially when they are under intense pressure. You are the type who is checking your phone for messages as soon as your eyes are open. That's just who you are. If you suffer from early morning insomnia, the best thing to do is to shift your sleep schedule so that you are getting ready for bed and relaxing by eight thirty p.m., with the idea of being in bed no later than ten. Then if you wake up at four thirty the next morning, you can just get up for the day. If you have cooled down your emotions and your body before bed, and then have six good hours of sleep that are aligned with your circadian rhythm, you are good to go.

Strong Sleepers

If you answered C to most of the questions, you are a strong sleeper. In Ayurvedic terms, you would be called a *kapaha* type. "Kapha" means water. Like water, you go with the flow. You are easygoing, methodical, and relaxed. As a strong sleeper, you may have a little trouble getting going in the morning, but once you get started, nothing gets in your way. You have tremendous endurance and an impeccable memory. You are steady, strong, loyal, and calm. You may have a tendency toward shyness, but you are an excellent listener. People come to you with their troubles and seek your advice. You are the glue for any contentious work situation, because you are a peacemaker at heart. In terms of your body's structure, you probably have larger bones, large eyes, smooth skin, and thick hair. You may gain weight more easily than your friends, but that's okay because a little extra weight looks good on you.

Strong sleepers tend to fall asleep relatively easily but are prone to oversleeping in the morning. They need the alarm and the snooze button, and on weekends, they joyfully sleep through half the morning. They need help getting out of bed and getting enough physical movement in their lives to shake off sleep. What's interesting about strong sleepers is that their bodies store lots of energy, so they don't really feel the effects of a bad night's sleep. Within a couple of hours of getting up, they are ready to think and work as though they slept perfectly. Strong sleepers who have been up all night rarely experience a troubled mood or reduced concentration from lack of sleep.

My patients who are strong sleepers rarely have trouble with insomnia. If they do, it's often because of physical complaints, such as congestion, snoring, and sleep apnea. When they can't sleep well, they gain weight. The irony is that when they gain weight, they have even more trouble sleeping. Also, they don't like getting up in the morning, no matter how much sleep they've had.

As a strong sleeper, here is what you need to do to better prepare for sleep and to get up on time:

Get moving in the morning. Strong sleepers need exercise to shake off their morning dullness. It doesn't need to be vigorous. A twenty-minute walk in the morning will power you up for the day and make sleeping easier that night.

Prevent congestion. I tell my strong sleepers that they should avoid snacking in the afternoon and evening, because any fatty foods and dairy products will increase their natural tendency toward congestion, and this is the primary problem they have with sleeping and snoring. I'll say more about diet in the next chapters, but sticking to a diet that reduces saturated fats and dairy and emphasizes vegetables and grains does wonders for a strong sleeper's constitution. With the right diet, you will be able to better control your weight and more easily avoid seasonal illnesses. You will also sleep much better.

Get rid of snoring. You may have been told that snoring is caused by an obstructed airway, jaw alignment, or the position of your tongue during sleep, but for most strong sleepers, snoring is really caused by congestion and the side effects of heartburn. If you can eat a light dinner by six p.m. and avoid all food afterward, you can lose a few pounds and your snoring will magically vanish. I've had couples come to me for help because one partner's snoring is keeping them from sleeping in the same room. What they don't know is that the timing of their evening meal has a lot to do with snoring. By eating a lighter, earlier dinner, many people can stop snoring within a few weeks, and couples can move back into the same bedroom. At the same time, they are sleeping better and feeling more refreshed in the morning.

Support your neck. Strong sleepers sometimes work to find a comfortable position to sleep, and they do better with a pillow that offers enough neck support. Elevating your head slightly may keep your airway free enough to breathe deeply during sleep.

A Note on Jet Lag

Most people have to deal with jet lag just a few times a year. But some people have work schedules that cause them to do quite a lot of travel, and they know all too well the heavy toll that jet lag takes on your body. Constant travel causes weight gain and insomnia and really does a number on your moods. A few months ago, a patient came to me for help trying to regulate her travel schedule. Leanne lives in Australia with her two young children, but she takes long flights almost every month. Her husband has recently been transferred to a bank in the US, and Leanne feels that she needs to bring the kids to him so that he can spend time with them. Meanwhile, she has aging parents in England, who insist that she visit them and bring the children. As a result of all of this travel, Leanne has gained almost eighty pounds in the past two years and she is miserable. She has tried to diet, but she keeps changing schedules and countries so often that she can't stick to any diet. She is forty-seven years old and doesn't want this weight gain to become permanent. Like a lot of frequent travelers, she has to deal with many different climates. In Australia, the winter months are warm, and when she goes to England, the climate is rainy and cold, and when she comes to the US, the winter months can be cold and dry. Her body is no longer experiencing the natural changes of seasons and the natural changes of light.

In addition to weight gain, she has begun to snore and to experience sleep apnea, which is not uncommon with abrupt weight gain. She says that she feels puffy and bloated in the mornings. Most people lose a pound or so while they sleep, but she has the opposite experience. This is a sign of inflammation in the body that can be a precursor to bigger health problems. Feeling bloated in the morning is one sign of inflammation, because the body is hanging on to water in the system as a way of cooling down the inflammation. You can test this at home. If you weigh more in the morning than you did the night before, you might be suffering from systemic inflammation. Loss of sleep from jet lag only exacerbates inflammation in the body. You need to find a way to reverse this trend.

My first concern for anyone with this problem is to reduce the amount of travel. For Leanne, this meant having her husband come visit her on some occasions and cutting down on her trips to England. If jet lag and continual travel are problems for your body, the first step is to cut down on the travel or at least leave a few more weeks between trips. In Ayurveda, we look for balance, and when your body is showing you that it is out of balance, you need to make changes. Leanne is a strong sleeper, so she needed a morning exercise routine to flush the bloating and congestion, and a light evening meal to reduce her heartburn. An evening dose of melatonin also allowed her to reduce inflammation, as it's one of the little-known benefits of this supplement. Once her schedule was in shape, she began to lose weight and sleep more soundly.

When you've done what you can to alter your travel schedule, the next step is to reduce the effects of jet lag on your body when you do travel. I relate personally to Leanne's struggle. Because I have extended family in India, but live in California, I also travel a lot. I visit conferences all over the world, and I understand that travel is a big part of life for many people. Once, people traveled between continents by ship, and during these trips they had weeks to adjust to the changing light and their new sleeping and eating schedule. But now we can hop on a plane and be in another part of the globe in twelve hours. The resulting shock to the body is sudden and very real. The good news is, there are a number of things you can do to minimize jet lag after long flights.

Lubricate. Your body needs a lot of help to adjust first to the dry, recirculated air on an airplane. I bring a bottle of oil drops with me in my carry-on bag. It can be very small and you don't need a lot of oil. You don't even need a special oil. Olive oil or sesame oil works fine. You want to put a couple of drops in each nasal passage near the start of the flight and after several hours of flying. Keeping the nasal canal moist will guard against the dry air and the buildup of mucus. I also put some oil in each ear canal. This is especially important on long flights, on which the air continues to dry your body.

Prevent overstimulation. Flights can be crowded and noisy. People wonder why babies cry on long flights. The answer is because it's too loud to do anything else. You have the constant drone of the engines, the white noise of the pressurization. There are people giving you instructions over a PA system, and these days there is often a television at every seat with a constant stream of video images. Every couple of hours, there is a beverage cart or food distribution. Who can sleep on these flights? But you need some sleep on long flights. I take eyeshades with me, and headphones to minimize noise. For extended flights, you want to get about five or six hours of good sleep before you land in order to minimize the jet lag. I tell light sleepers not to go anywhere without eyeshades and noise-canceling headphones. Without them, you will be too frazzled when you land to get your bearings. Even if you don't sleep, the act of minimizing stimulation relaxes your body and helps you deal with any of the stressors of the trip.

Use melatonin. Flying from west to east is the most difficult because you are jumping your body ahead in time. The morning comes earlier and night falls before you are ready to sleep. For example, if you are flying from New York to Paris, you will be trying to fall asleep at ten p.m., when your body thinks it's still four in the afternoon. You want to be taking melatonin at four p.m. for a couple of days before the flight. It won't make you sleepy, but it will get some melatonin into your system at the right local time, so that when you do land, your body will be primed to try and sleep at that time.

Fast during the flight. If possible, you want to reduce or eliminate eating on the airplane, particularly if flying east. Food is a powerful signal for the body to help it adjust to a new environment, and it may be the most powerful tool for avoiding jet lag. When you fast, your metabolism slows down as though it's nighttime. This helps you to relax or sleep and it helps advance your circadian rhythm to the new time. When you get off the plane and eat again in daylight, your body will know that a new day has started. It will have taken the first steps for entraining your system to local time.

Take triphala. This is an Ayurvedic herb that helps to regulate the bowels. Many of my patients take a one-thousand-milligram tablet in the evenings because it is a great digestive aid. It also helps to lubricate the bowels. It's not a laxative, but it frequently has the effect of softening stools and keeping things moving. If you suffer from constipation after long flights, this supplement is a must. Take it as the plane lands, or just after, and you will prevent indigestion and constipation as you adjust to the new time zone.

Live on local time. Do this as soon as you land. If you fly east, you might land in the early morning hours of the local time after minimal sleep in flight. It's important to avoid napping to make up for any sleep deprivation on the flight. Stay awake until the local time dictates sleep. It's also important to eat food at the local time. If it's lunchtime where you are landing, eat lunch. This helps your body and your hypothalamus to reset the internal clock to a new twenty-four-hour schedule.

Get light. During the day in your new time zone, you want to get outside as much as possible to be exposed to the natural sunlight. This more than anything will help your body reset its rhythm, and it will help not only your sleep schedule but your digestive schedule as well. This is the biggest predictor of jet lag, not getting enough natural light.

Prepare for sleep. On those first few nights in a new location, you want to put extra care into your evening routine. Minimize the emails and electronic stimulation. Give yourself an oil massage and take a soothing bath before you go to bed. By doing this, you are teaching your body its new bedtime.

Once you make the commitment to get a better night's sleep, you will notice changes in every part of your life. In a way, you are eliminating the natural jet lag you have imposed on your schedule. As a result, you will notice a change in your digestion and your level of focus during the day—even your skin will glow with a few nights of good rest. Many of my patients lose weight when they get the proper rest, and you may notice the same. In the next chapter, I'll help you keep this healthy glow by outlining a new way of eating. When you eat better, you will lose even more weight. You will sleep better, too.

6 You Are When You Eat

See if this sounds familiar. You wake up in the morning, thinking that today is the day you are going to eat less and start losing weight. You aren't hungry in the morning, so you skip breakfast. (The majority of people who need to lose more than twenty pounds aren't hungry in the first three hours after they wake up. This is often because they ate too much the night before.) You go to work, and by lunchtime you do feel hungry, but you have a little salad at your desk or a small sandwich and you congratulate yourself on your willpower. By three or four p.m., that willpower is gone. You are hungry, cranky, and fatigued, and you are now craving something to take the edge off. Maybe you go to the vending machine. Maybe you get a giant sweetened coffee at Starbucks. You know that neither of these is the healthiest choice, but at least you can power through the rest of the day. After work, you are starving and exhausted. This is when the real trouble begins. You reach for a glass of wine while you cook dinner or a snack while you wait for your takeout to arrive. If you go to a restaurant, you eat all of the bread before the meal even arrives. After dinner, you can't quit eating, even though you feel full. Your blood glucose levels and insulin levels are spiking, but you crave something sweet. You reach for dessert or continue to snack until bedtime. Unfortunately, food consumed late in the day, when your digestive tract is preparing for its rest cycle, won't be digested. You may have acid reflux, congestion, or intestinal cramping. So your sleep is less restful, and you wake up feeling full from the night before. Starving yourself in the first half of the day and bingeing in the second half of the day will sabotage your weight-loss efforts.

If you have fallen into this pattern, you aren't alone. Nearly every one of my overweight patients tells this same story. In fact, almost everyone who follows the standard American diet consumes about half of their daily calories after four p.m. In one observational study, researchers asked 156 participants to take a photo of every meal or beverage before consuming it every day for three weeks. (It's actually a great idea for logging your food intake.) What they discovered was that about half of respondents took in 75 percent of their daily calories in the afternoon and evening.[1] They also found that people further delay meals on weekends, worsening the jet lag effect on the body.

In the late afternoon and evening, the body is far less able to metabolize calories, sugar, and fat. People think that skipping breakfast or lunch is saving calories for later, but this isn't true. Your body isn't a simple machine that consumes a set number of calories every twenty-four hours. Your digestive tract has its own circadian rhythm. It is ready to go in the morning and burns most efficiently at noon. After two p.m., it becomes progressively more sluggish. People with a slower metabolism who struggle to lose weight have almost no ability to digest any food after seven p.m. And eating your biggest meal in the evening causes a host of metabolic changes that make weight loss impossible.

Meal Timing Matters

Very few dietary guidelines talk about the timing of meals, and yet obesity researchers are beginning to say that meal timing is the missing link in weight management. Your body's ability to handle a large infusion of calories—its glucose tolerance—is higher in the morning than it is later in the day. And insulin sensitivity is also cyclical, with insulin sensitivity higher in the morning and lower in the evening.[2] So if you have some carbohydrates in the morning, say a little bit of oatmeal or a piece of whole-grain toast, your body's blood sugar won't spike as much as if you ate those same things in the evening. Of course, if you've been starving yourself all day, you are probably reaching for high-fat simple sugars, such as corn chips, buttered bread, and sticky desserts. On this fuel, your body's blood sugar will rise sharply and stay high for several hours just because you are eating these things after dark. What's worse, the body is also preparing for lipogenesis, or fat storage, at the end of the day. Consuming most of your calories in the evening causes that giant meal to be stored as fat.

If you think I'm describing the typical diet as a slow-motion disaster for your metabolism, you are right. In 2014, researchers published a study about the long-term effects of eating a large dinner. Researchers asked 1,245 people of normal weight with no metabolic issues to fill out a three-day food diary and to take some blood tests. Six years later, those same people came in for testing. Those who reported in their initial diaries that they ate more than half of their daily calories in the evening meal were twice as likely to have developed obesity or some other metabolic issue in those six years.[3] Moving your largest meal to the middle of the day, even if you don't change anything else about your eating habits, could change your health for years to come.

Can you lose weight by simply eating your largest meal earlier in the day? Yes. Researchers asked overweight and obese women with metabolic syndrome to adhere to reduced-calorie diets. One group ate their largest meal at breakfast, with a smaller meal at lunch and a very small meal at dinner. The other group reversed the order, eating just two hundred calories at breakfast, five hundred at lunch, and seven hundred at dinner. After twelve weeks, those who ate their largest meal at breakfast lost more weight and had a smaller waist circumference. They also showed improvements in insulin sensitivity, fasting glucose, and triglyceride levels. More important, they reported less hunger than those who ate their largest meal at dinner.[4]

Now, I would never suggest you eat your largest meal at breakfast—you'll see why in a minute—but I do want you to give up on the idea of skipping breakfast altogether. Some research suggests that dieters can keep their weight loss going for at least seven months by eating breakfast because it lowers hunger and food cravings.[5] If you aren't very hungry in the morning, you can think of it as a snack or a primer meal. It's a little something to get your metabolism fired up first thing. When you've been able to trim down that evening meal, you'll find that you wake up with more natural hunger in the morning—an important sign that your metabolism is normalizing. Many people tell themselves that they should eat when they are hungry, but you are forgetting that you've trained your body to be hungry at the wrong time of day. You've got to turn this around. By eating in the morning, you are training your body to be a little bit hungry in the morning, to break the body's natural fast when those calories can do your body the most good—and the least damage. Changing your mealtimes by adding a little bit of breakfast and a substantial lunch eliminates that shaking feeling of hunger late in the day because it teaches your hunger hormones to do what nature intended.

The Hunger Hormones

Your eating habits train your body to produce certain hormones associated with eating, including ghrelin and leptin. Ghrelin is produced in the stomach, and it's the only hormone the body makes to stimulate appetite. Whenever you eat, you are training your body to anticipate food at this time, and to make you very hungry at the time when you normally eat. You might notice this phenomenon during the change to daylight savings time, when the clocks change but your stomach doesn't. That's the body's ghrelin production at work. It doesn't know that the time of day has changed, and it can take several days or weeks for your body to fully adjust. Ghrelin rises sharply before you eat and decreases after you eat for several hours. It also increases sharply just before nightfall and then decreases just as quickly after dark and stays low until morning. Even though your stomach is (or should be) empty at night, you don't feel hungry or anxious about getting food. This is the natural way that the body regulates eating. Ghrelin also tells your body to prepare to store fat. What does this have to do with what time you eat? Well, if you are routinely eating your largest meal after dark, you are training your body to produce ghrelin when it would normally be suppressed. You are also telling the body to prepare to store abdominal fat, and you are doing this right before you lie down for eight hours and put your brain and organs into sleep mode.

Leptin works the opposite way. It is released by your fat cells to tell the body that you've had enough to eat. Leptin levels will be low before you eat but higher afterward. If you have enough stored fat, those cells should be releasing leptin into the blood as a natural appetite suppressant. The discovery of leptin at first seemed like a breakthrough in the study of obesity. Perhaps people who gain weight don't produce enough leptin or could be given a leptin supplement! It turns out, though, that this isn't quite true. People with adequate fat stores do produce quite a lot of leptin, but for some reason, their appetites don't respond to the high leptin levels in their blood. Your body can develop leptin resistance, which is sort of like insulin resistance. Your brain stops receiving leptin signals when you don't get enough sleep or when you eat at a time when your body would otherwise be sleeping.

Like ghrelin, leptin follows a circadian rhythm. Lower during daylight hours, it rises naturally at night to tell your brain that you don't need to be eating. In fact, people who are obese can produce more than twice the amount of leptin at night as people of normal weight, and these high levels of circulating leptin contribute to leptin resistance. Eating a big meal at night also increases circulating leptin levels at a time when they would normally rise.

That's what was happening with Ron, a thirty-one-year-old advisor at a wealth management company. His nutritionist had told him to make sure to eat breakfast, but that big morning meal was making him feel sick. He was forcing himself to eat a large breakfast even though he wasn't hungry in the morning. What he wasn't telling his nutritionist was that he often skipped lunch. Then he would be ravenous at dinner and overeat, which was contributing to his insomnia. A big breakfast does nothing for you if you are going to be fasting for the next twelve hours. Your body needs fuel during daylight hours. It needs to fast after dark. Ron thought that his painful hunger was a signal from his body that he could eat and drink whatever he wanted—soda, bread, meat, dessert. If you are too hungry, everything sounds good, and you give yourself permission to binge. This may satisfy cravings in the moment, but it is a metabolic disaster. At night, your stomach empties at half the rate it does by day. Food consumed at night won't be easily digested. By morning, you will still be queasy from that big dinner and your leptin levels that are still too high. In fact, fasting during part of the day contributes to leptin resistance. By contrast, your ghrelin levels will be too low to feel any hunger. The pattern you've established for your body works completely against your natural rhythm, feeding fat stores while starving your body of essential nutrients in the first half of the day, when it needs them most.

How can you break this cycle? You start by restricting calories at night. Then you can have a modest breakfast of a small smoothie or protein shake or some whole grains or even eggs with vegetables. The goal is to push your natural window of hunger to the noon hour, when you can safely eat more substantial food. Thousands of my patients have used this new eating pattern to normalize their hormone balance and improve their blood work. It also supercharges their willpower because they no longer experience that desperate hunger.

You may need to experiment a little to get used to eating less at night. Ron eventually learned to eat a light dinner at six p.m., but for the first few weeks, he would eat a little piece of fruit or drink a cup of almond milk at eight p.m. to combat that habituated hunger. Remember that you have to train your brain to signal hunger at the right time, and tried-and-true methods for reducing hunger, such as drinking hot water with lemon, can help you manage hunger signals when they pop up at the wrong time.

Your Gut's Clock

The master circadian clock is located in your brain, but many organs and tissues (including your fat tissues) have their own circadian rhythm. Eating is a primary means by which these peripheral clocks, the ones in your gut and liver, decide what time of day it is. They need to set their own clocks by when you eat so they can decide when to release digestive enzymes, divide and repair cells, absorb nutrients, and move waste along. This happens on a cellular level, using those molecular clocks, or clock genes, inside their cells. But these peripheral clocks are also constantly talking to the brain's master clock, the SCN, to give information about the body's activity. While the brain is using the light/dark cycle and your sleep patterns to set some functions, the gut is using feeding times to try and set its own twenty-four-hour cycle. When these are misaligned, you get the classic symptoms of jet lag, even if you never got on an airplane. Not only do you have irritation and inflammation in the bowels, but you have persistent fatigue and brain fog. You can also suffer from mood disorders and difficulty with concentration or dealing with stress, because the gut is the primary driver of serotonin production. That's what people mean when they talk about the brain-gut axis. Now we know that a big part of that brain-gut axis is related to the circadian clocks all over your body.

In addition, the microbiome in your gut has its own circadian rhythm with some bacteria more prevalent in the daylight hours and others at night. When your eating patterns conflict with the master clock, the whole balance of microbes in your gut is disrupted. You can disrupt the microbiome by eating a high-fat diet, by eating late at night, or by eating for too many hours during the day. This population of microbes can get so disrupted by odd feeding times that they lose their way. The populations of bacteria stop operating on a diurnal rhythm, and this leads to things like inflammation and bloating, but also glucose intolerance and obesity.[6]

So your improvised diet, the one that has you fasting or eating lightly during the first part of the day and bingeing on food at or after sunset, can cause a lot of disruption in your body, including acid reflux, ulcers, irritable bowel syndrome, insulin resistance, and weight gain. But don't despair, because there is one substance that the body produces that synchronizes all of these clocks and improves your digestion: melatonin. Falling asleep on time gets melatonin flowing from the brain before midnight. Perhaps I harped on it in chapter four, but this is why: going to bed every night at the right time teaches your body to release a flood of melatonin into your system that will coordinate the clocks in your gut to the clock in your brain. And it turns out that any melatonin will do. So if you are taking melatonin supplements to help regulate your sleep, these are also going to have a protective effect on your digestion. It helps calm the digestive tract and regulates bowel movements. In one study, women with IBS were given an eight-week course of melatonin (three-milligram doses each night) and showed a marked improvement in their symptoms.[7] Melatonin is also a powerful antioxidant in its own right and may have a protective effect on the liver. Boosting your body's nighttime melatonin production will do more for you than any standard weight-loss regimen.

Eating at the Right Time

Changing your diet means changing your pattern of eating to put your body in sync with the circadian rhythm. We will get to clean eating and specific food choices for your body type in the next chapter, but first, you have to eat your meals at the right time.

Tracking Your Intake

In order to know how to change your eating schedule, you have to know what it is. For the next few days, keep a notebook handy to write down everything you eat and specifically at what time of day. You can use your phone to take a picture of your food, which includes every beverage and snack. I'm not interested in the specific number of calories you think you have consumed. I'm more interested in the relationship between one meal and the next. By tracking your pattern of eating, rather than the calories, you can get an idea of when your body expects food from day to day. Be prepared to answer these questions as you analyze your food log:

- **What time do you eat each meal?** Do you eat at the same time each day?

- **What is the size of each meal?** You can estimate calories if you want, but what you're really looking for is which meal is the largest and at what time of day are you eating it.

- **How many times a day do you snack?** Keep track of everything you eat between meals. This means every sweetened coffee, every small bag of chips, and every piece of candy. You are looking for the pattern of snacking.

- **How many hours a day do you avoid food?** How long is the fast between meals and between dinner and breakfast? This tells you how much rest you give your digestive tract every day.

- **How colorful are your meals?** How many different vegetables, nuts, grains, and other natural foods are you consuming with each meal? Your goal is to load up with as many naturally made foods as possible. (More on this in the next chapter.)

- **How many complex tastes do you enjoy at each meal?** There are six dominant tastes: sweet, sour, salty, bitter, pungent, and astringent.

But the modern diet focuses primarily on sweet and salty tastes, so that's what you crave. You need to expand your palate to include other tastes and break the lock on your taste buds that steers you toward junk food. You need pungent foods, such as spices. You need astringent foods like citrus, vinegar, and celery. You need bitter foods, including greens, sprouts, and melons. These aren't just good fiber; they actually cleanse the palate and help you make better food choices. The ginger drink on page 127 contains all of these tastes, which is why it's such a simple way to eliminate food cravings.

By tracking your food intake, you are bringing mindfulness to your diet. You can't change what you are eating until you know what you are eating—and when you are eating it. We talk about meditation bringing the heart and mind together. A food journal brings your stomach and mind together. It's the first step toward knowing that food isn't just what tastes good in the moment: it's what feels good in the hours after you eat.

Eat at the same time every day. Many people with busy schedules have no idea from day to day when they will eat. I worked with someone in sales who said that she had to meet with clients several days a week and often took them to lunch. Never mind that she was eating spicy food one day and heavy pastas the next, her lunch meetings ranged from eleven thirty a.m. one day to two p.m. the next. This inconsistency is a disaster for your digestive tract. She needed a schedule. Your body releases digestive enzymes and hormones in anticipation of mealtimes, and it needs to be able to predict when to do this while you are training your body to be hungry at the right time. If you have to, set an alarm in order to eat at the same time every day for at least a week. Especially on the weekends, when people tend to eat later. You want to avoid this social jet lag effect on your mealtimes. This is how you train your body to expect foods at the right time. If you take the challenge to flow your work schedule around your eating schedule, your body will thank you. My patient did this, and at the end of the week, she couldn't believe how much better she felt. She began to encourage her clients to eat at restaurants where she could get more vegetables and cleaner foods, and they loved it. On other days, she packed a lunch of soup and salad and felt fantastic.

Enjoy your biggest meal in the middle of the day. This helps reset your circadian rhythm and your digestion. You want to be eating most of your calories in the middle of the day, when your digestion is going full force. It doesn't have to be exactly at noon. According to the Ayurvedic day, that pitta period of peak digestion is between ten a.m. and two p.m. Still, I advocate for setting a time for your lunch that's as close to noon as possible and sticking with it every day. The evening meal should be more like a snack, something that takes the edge off of your hunger before night falls. I tell people to do this before we talk about revitalizing the breakfast meal, because most people aren't hungry in the morning until they remove the big evening meal.

Prepare a nutrient-rich breakfast. Many people think that breakfast can be enormous, while others skip breakfast entirely. Both of these are mistaken. In Ayurvedic terms, the earliest part of the day, between six and ten a.m., is considered a time when the nervous system and digestion are still warming up. It's a time for the first bowel movement as your digestive tract purges yesterday's waste. And it's not a time to load up the stomach with flours, sugars, and animal fats. A smoothie with lots of greens and fruits or a little bit of oatmeal usually does the trick. If you need more grounding, though, you can have eggs with vegetables or potatoes. The energy from these foods needs to power you only until lunch. If you don't have a strong appetite in the morning, there's no need to pack in food now.

Set your intention to avoid snacks. After each meal, you want to make your intention to eat nothing until the next meal. Gentle fasting in this way is good for you and it's good for your body, allowing it to rest, optimize the gut flora, and let your leptin and ghrelin levels stabilize. Reach for water, even hot water or herbal tea, to take away hunger while you are training your body to expect food at the right time. This is especially useful after dinner. You know that your body is working to suppress your hunger until morning. Let it do its work. You don't need to eat at night. Remember: you are training your body and, with time, this will be natural for you.

Eat Fewer Meals

One of the things I ask people to pay attention to in their food log is how many hours a day they spend eating versus fasting. If you consume some calories first thing in the morning (even if it's just coffee with cream alongside a piece of toast), you are firing up the digestion right at six or seven a.m. Then if you are snacking during the day and finishing your dinner at nine p.m., that's fifteen hours of the day you spend working your digestive tract. If you are snacking right up until bedtime, your body isn't going to have much rest. I tell people that they want about ten to twelve hours of fasting time at night, and they want at least three hours between meals with no food. This kind of gentle fasting is good for your body. It's not necessarily about eating less. Rather, it is about eating less often.

Many of the people I meet in my practice are simply eating for too many hours of the day. They drink sweetened coffee first thing in the morning and off and on until noon. They graze throughout the afternoon, eat a big meal in the evening, and then continue to snack until bedtime. Of course, they think they are taking in too many calories, and that may well be the case, but more important than calories or fat intake, they are simply eating too often.

Having set mealtimes is an easy way to combat grazing tendencies and mindless eating. If you stick to a schedule and know when you are going to eat and when you are not, in time, so will your body. Earlier, I mentioned a study in which researchers asked participants to take pictures of their food for three weeks. One of the findings was that for most people there is no such thing as three square meals in one day. Instead, people eat and drink between four and fifteen times a day. While there was an average interval of three hours between episodes of eating, about half of the respondents were eating off and on for about fifteen hours a day. Researchers recruited eight participants to eat for an intervention, in which they were asked to restrict the daily window for taking in food to ten or eleven hours. They were not asked to make any other dietary changes. Those who restricted the number of hours in which they ate lost an average of seven pounds in sixteen weeks and were able to keep that weight off for a year.[8]

In the evening, your body should be preparing itself for a natural fasting phase in which the cells lining the esophagus, stomach, and intestines can carry out normal repairs. The cells are clearing out debris, replicating if they need to. During this time, your digestion slows and bowel movements are suppressed. According to Ayurvedic wisdom, your body needs at least ten hours of fasting time at night to carry out these repairs. It also needs time in between meals to rest and kindle feelings of hunger. The pleasant feeling of hunger in anticipation of a meal is something people enjoy too rarely. Of course, you want to stop eating a late dinner, and you want to avoid eating anything after dinner. You also want to begin to train yourself to avoid eating between meals. This is far more important than training yourself to eat low-fat foods or avoiding certain categories of food.

Why Diets Fail

You might expect me to say that diets don't work. Instead, I'm going to say the opposite. The trouble with most weight-loss regimens is that they do work. At least temporarily. This is their seductive power. If you restrict calories, or highlight proteins while avoiding carbohydrates, or eat smaller meals or avoid certain foods altogether, you will probably lose a few pounds. Yet most diets don't consider meal timing. Some of them encourage you to eat your largest meal at night, which will eventually cause your weight loss to plateau. Worse, diets assume that your biggest problem is your weight gain. The problem you need to solve first is a confused circadian rhythm that's wreaking havoc with your gut microbes, your hormones, and your sleep. Weight gain is just the result of those disruptions.

Also, not every diet will work for every type of body. People with a naturally slower metabolism will do fine on a diet that sharply restricts calories. They will do fine on a diet that is vegetarian or even vegan.

Someone with a strong metabolism will go crazy on a low-calorie diet, and will feel so hungry that they can't concentrate. I'll talk about different body types and the best diet for each type in the next chapter.

Working Around Your Work Schedule

If you're like most people, your work schedule controls everything from the time you get up in the morning, to your mealtimes, and whether you have time for exercise. You may be thinking right now that you can't take all of this good advice because of your work schedule.

Many people don't realize that their work schedule is the main culprit in their digestive issues. That was the case with Wendy, who had come to see me because of gastrointestinal issues and weight gain. She was working the evening shift at an Italian restaurant, so she didn't eat dinner until her shift ended at ten p.m. At that point, the entire waitstaff would sit down to eat pasta and drink wine together. It was a way of discharging stress after a busy and stressful workday. As a result, she was tossing and turning all night with an upset stomach. She was getting up late with a queasy stomach and fasting for much of the day to save those calories for the end of the day.

Wendy's coworkers were in their twenties, a time when the circadian rhythm is strong and you will feel comparatively fewer effects from a poor schedule. But Wendy was in her late forties, and she had gained twenty pounds in the previous year. She needed to find a way to eat well despite her demanding schedule.

Lots of people need to work until later in the day, but delaying that final meal until you get home isn't healthy. Other patients of mine say that they have lunch meetings on some days that are dictated by clients or by management decisions. You have to be careful under these different conditions. Here are a few things you can do to support your circadian rhythm even if your work schedule poses some of these challenges:

If you routinely work late . . . Make sure you get up early and do your morning exercise before breakfast (more on this in chapter eight). Eat breakfast and a hearty lunch. At six p.m., step away from your desk to a quiet spot and eat a light dinner. It may mean packing two meals, but keeping groceries like fruit and precut vegetables at the office can help you find time for a light evening meal. Avoid delaying your dinner until you get home if you work late every night. Instead, plan for your dinner while you are still at work. If you work late occasionally, you can delay dinner once in a while, but make sure that evening meal is especially small. Stay away from technology after work and try to fit in some meditative activity. In the hours after work, you want to be actively relaxing without snacking so that you can get to bed on time.

If your lunchtime varies because of meetings . . . One patient of mine had regular conference calls with important clients that overlapped with her lunchtime. She learned to step away from the calls for about fifteen minutes and let her coworkers stand in for her, so that she could eat and come back. You may not have this flexibility. If your lunchtimes have to fluctuate, then be careful about what you eat. No heavy carbs, and nothing starchy when you do eat. If you have the ability to set your lunch schedule—for example, deciding what time to make a lunch reservation—try to pick a time as close to noon as possible. Regardless, you should keep the rest of your schedule intact, including sleep and exercise times.

If you work the late shift . . . The biggest mistakes people make during true shift work is they eat at night to stay awake. Then they eat at the end of their shift as a simulated breakfast, then they nap and eat throughout the day. Instead, you should fast as much as possible during the night hours. You can drink lightly sweetened green tea if you need help staying awake. When you get home in the very early morning, sleep first, while it's still dark, exercise, and then eat. You want to keep as close to the natural circadian rhythm as you can.

Avoiding Food Cravings

One of my patients was a woman who works for a large nonprofit. She routinely hosted large fund-raising parties, and she told me that the holiday and fund-raising seasons were difficult for her because of the delicious food served at parties. She felt that she just couldn't resist the rich appetizers floating around and the huge buffet meals served, and they contained all of the foods she knew she was not supposed to eat. The parties themselves were lovely, but she would always wake up the morning after feeling bloated and weary. If she had to attend several parties in a one- or two-week period (as many people do during holiday seasons), she would find herself craving bad foods all the time and gaining weight. She asked me how she could stop this cycle.

When I talk to people about changing their diets, I always emphasize the difference between what tastes good on the tongue and what tastes good to the body. Many foods high in salt and sugar taste good while they are in your mouth but make you feel sick after you've eaten them. One answer is to get as many tastes into each meal as you can. Junk food isn't just high in fat, it's high in sugar and salt. It teaches you to crave these things in the future. By adding as many of the six dominant tastes to your diet (sweet, sour, salty, bitter, pungent, astringent) as possible, you can teach your tongue to like different things. It will no longer crave just two tastes. You will be amazed at how quickly your taste for junk food disappears when you have other tastes in your diet. In the meantime, we have to trick your tongue, which is surprisingly easy to do.

If you have ever gone to a multiday workshop at the Chopra Center, you will know that they serve a little liquid in a shot glass before each meal. It's a potent mixture of ginger, lemon, salt, pepper, and honey. I've taught at a lot of these events, and I can report that about half of the people resist this drink, especially after they try it. They pucker their mouths, grimace, and say out loud how much they hate it. Fair enough. Nobody ever said that personal growth had to be tasty from the very first sip. What's interesting to me is that, at the end of the workshop, those same people always ask for the recipe for this drink. Not only have they come to like the taste, but they begin to feel that they need it. They have noticed that after drinking this, they enjoy the taste of their food more, they eat less, and they don't have the same food cravings that they did before they came to the workshop. How can this be?

This drink offers five ingredients that contain all of the dominant tastes. In one tiny serving, you are getting something that is simultaneously sweet, sour, salty, bitter, pungent, and astringent. Because it is so complex, it awakens the taste buds. In one gulp, you have satisfied your taste buds before you put the first bite of food into your mouth. Ideally, this is what an appetizer in a restaurant would do. But what do you usually eat when you go to a restaurant? Bread with butter or oil. Or something heavy, flour-based, or fried. This is not waking up any taste buds. Instead, it puts them to sleep. At cocktail parties, you are probably being served platters of appetizers that are oily, heavy, and salty. You are getting nothing spicy, bitter, or pungent. And if you are drinking a cocktail, that also puts your taste buds to sleep. How can your body ever know that it has taken in enough satisfying food?

I tell people to try this ginger beverage before each meal. You don't need a lot of it. Just drink ¼ teaspoon of this before each meal to get the taste, and you will see how your eating changes and how your craving for certain foods diminishes. Many people tell me that after they use this drink for a week or so, they can taste their food in a completely different way. Formerly bland foods, such as vegetables, take on rich subtle notes. And formerly tasty foods, such as candy, start to have a chemical tinge to them. One patient told me that he had one bite of his favorite candy bar and it was absolutely disgusting to him.

Tasty Ginger Drink

2 teaspoons freshly grated ginger
4–5 teaspoons honey or Sucanat or date sugar
¼ teaspoon Himalayan salt
¼ teaspoon black pepper
2 teaspoons lemon juice

In a glass container, mix all the ingredients well. Keep it in the glass container in the refrigerator, where it will stay fresh for about 1 month. Drink ¼ teaspoon of this mixture first thing in the morning or before meals. You can dilute it with some hot or warm water, if the taste is too intense at first.

How well does this work? Well, I gave this to my patient who had to host all of those fund-raising dinners. I also advised her that she should make her own dinner at home, a very light meal of steamed vegetables and a little bit of rice. I then told her that she should take a tiny taste of the ginger drink, eat lightly before going to the party, and then see how she felt about the food there. She came back to me and said that she didn't eat a single appetizer at any of the parties she attended. In fact, she told me that the food didn't appeal to her at all.

Meal timing truly is the missing link to weight loss and optimal health. By eating a healthy breakfast and a substantial lunch, you can strengthen the brain's circadian rhythm and protect your intestinal tract and your gut microbiome. You can normalize your weight without intensely restrictive diet regimens. But it's also important to eat the foods that are right for your body. In part, you have to eat the foods that are right for *any* human body, meaning foods that are healthy and nourishing. Fast foods and prepackaged foods are not an American phenomenon. They are a global phenomenon. When I go back to big cities in India, I sometimes think I might as well be in a suburb of Chicago because there are so many fast-food restaurants and so many prepackaged foods in the aisles of the markets. But you can better resist these unhealthy choices with a good meal schedule in place.

Your second task is to find the diet that's right for you. Not every diet works for every type of body. In the next chapter, I'll talk about how you can figure out what kind of body you have and what kinds of foods will best support your optimal health.

7 The Right Diet for You

The problem with a typical weight-loss regimen is that it assumes that every body has the same relationship with food, and that everyone needs to reduce calories in order to lose weight and feel better. Of course, we all know this is not the case: if you look around a room, you can see that everyone has a different body and different eating challenges.

Bruce came to see me recently after an unhappy encounter with his doctor. His most recent blood work and physical showed an increase in cholesterol, fasting glucose, and blood pressure. He was not high enough in any one of those categories to require medication, but the doctor warned that it was only a matter of time before he would need multiple medications to keep him from having a heart attack or developing metabolic syndrome. "Lose some weight" was the doctor's parting shot. Bruce was frustrated to hear this. At 239 pounds, he *knew* he wanted to lose some weight. At nearly six feet tall, he had carried extra weight well and, until recently, had always looked beefy rather than fat. Bruce is a self-described foodie and a workaholic. He runs a small business that offers nutrition-based education programs.

He travels around the country teaching people to eat better. And he runs a personal blog that reviews restaurants based on the nutritional quality of their food. He is a devotee of healthy, clean eating, so why was his weight careening out of control?

From a chronobiological perspective, Bruce had a couple of problems. The first was that he had abused his body's naturally strong metabolism. If you are tall, muscular, or athletic, you probably also have a strong metabolism. That's a good thing when you are young, because it allows you to eat more than other people do without gaining much weight. It's tempting to begin to build your life around food and food-related events. People like Bruce often feel that eating well negates the effects of eating too much, but it doesn't, particularly when those meals come from restaurant food with the enormous portions and extra doses of salt and sugar. A lot of restaurant food contains ingredients that aren't so healthy. It's easy to overeat, even if you never touch fast food.

Turning forty didn't help, either. Middle age is a critical point for so many food lovers. As your metabolism naturally slows, you may find that your old eating habits don't work anymore. Your diet hasn't changed, but your body has. Lack of sleep and little time for exercise only exacerbate the situation. Some people find themselves careening suddenly into obesity without the right diet plan. But which plan is right for you? It's hard to know what to change unless you know how your body works. I offer patients a body type quiz to help them understand their lifelong relationship to food. This quiz also focuses on the one thing that most diets don't. It asks about your digestion, which is the key to your body type. After all, in Ayurveda, we know that your weight isn't only about what you eat. It's also about how you digest.

You can discover your own metabolic type by answering

these questions. Again, there are no "right" answers, and it's okay if an answer doesn't suit you perfectly. Just try to find the answer that best fits your situation most of the time.

1. When I was a teenager, my parents' biggest complaint about my food intake was that:
a. I was a picky eater.
b. I was hungry almost immediately after a meal ended.
c. I was eating too many sweets and fatty foods.

2. In my teen years, my body frame could be described as:
a. Thin or small.
b. Muscular or tall.
c. Stocky, regardless of height.

3. When I go to an unfamiliar restaurant, my number one concern is:
a. Will they have anything I can eat?
b. Will it live up to the hype?
c. Will they have any comfort foods?

4. When I'm forced to skip a meal, or forget to eat, I feel:
a. Scattered or light-headed, but not always hungry.
b. Ravenous and irritable.
c. Fine. Sometimes I feel good.

5. In the mornings, I am:
a. Rarely hungry.
b. Usually hungry.
c. Still full from my evening meal the night before.

6. If I have an overwhelming day at work, I'm likely to do what at lunch:

a. Forget to eat, or grab something quick, even if it's unhealthy or unsatisfying.
b. Make sure I eat a decent meal, even if it's takeout food.
c. Eat my regular lunch, but then go looking for something sweet as a reward.

7. When I do skip lunch, it's because:

a. I forget to eat or I'm not feeling hungry yet.
b. I'm exercising or finishing work under deadline.
c. I rarely skip meals. I like taking the time to eat.

8. In terms of making a healthy dinner, I:

a. Often don't have the energy to cook in the evenings.
b. Really want to sit down and enjoy it, even if it's late at night.
c. At the end of the day, I want a treat. I don't want to think about dieting.

9. When I try diets, I struggle because:

a. I'm very motivated for the first few weeks, and then life gets in the way.
b. By the end of the day of dieting, I'm so hungry that I give in and start snacking.
c. The food cravings never go away. It's hard to give up my favorite foods forever.

10. My biggest challenge in starting a new eating regimen is:

a. I'm overwhelmed by the details, the shopping, planning, and cooking new things.

b. I'm afraid I'm going to be hungry all the time.
c. It's difficult to resist the temptation to eat things I shouldn't.

11. What are your typical results from past diets?

a. I lose five to seven pounds and feel good, but not good enough to work that hard.
b. I am vigilant about reaching a goal weight, no matter how much it hurts.
c. My body just doesn't cooperate. Waiting for that number to change on the scale is like torture.

12. When I think about how I'd like to change my diet, my biggest problem is:

a. I obsess about calories sometimes, but other times I forget to count them.
b. I know I shouldn't snack all day long, but I can't think when I'm hungry.
c. I turn to food when I feel stress or depression.

13. When I think about losing weight, I think:

a. If I change one or two habits, my weight would go down. It always has.
b. I could lose weight if I would exercise more.
c. My weight has been a problem for a long time, and I wonder if that will ever change.

14. When I travel across time zones, my biggest digestive issue is:

a. Nausea. I'm not hungry at the new mealtimes.
b. Constipation.
c. I feel bloated and retain water.

15. I snack between meals primarily because:

a. I'm a little bored or I forgot to eat a meal.
b. I'm hungry all of the time.
c. I need comfort or a distraction from what I'm supposed to be doing.

16. I would say my bowel movements are:

a. Sometimes dry or hard, and sometimes I have constipation.
b. Like clockwork.
c. Sometimes smooth and well formed, but sometimes loose and watery.

TOTAL: A: ____ B: ____ C: ____

If you answered mostly A, you have what I call a variable (vata) metabolism. If you answered B to most of the questions, I would say you have a strong (pitta) metabolism. And if you answered mostly C, you have a slow (kapha) metabolism. It is neither right nor wrong to be any one way, and it's okay if your eating type doesn't correspond to your sleeping type. This is just a reflection of how your body reacts to food and how your attitudes about eating and weight gain have been shaped in part by your body's reaction to food.

Variable Metabolism

If you answered A to most of the questions, you have a variable metabolism. Like the light sleepers, you are called a "vata," which means air. While it's true that light sleepers often have a variable metabolism, it's possible that you have one kind of sleeping style and a different eating style. People with a variable metabolism never quite know when they will be hungry and, as a result, their mealtimes tend to fluctuate from day to day.

In childhood, you might have been a picky eater, or someone who sat at the table eating a few bites here and there and never quite finishing. You may have been chatty while eating, or you were so interested in the conversation swirling around you that you forgot to eat. Or you were daydreaming and too distracted to focus much on food. You can probably remember times as a child when you just didn't feel hungry at mealtimes. You probably had a small stature and small bones, particularly at the wrists and ankles. As a child, you may have had times when you were running around, eager to meet with friends, but also needed time alone to recharge.

As a teenager, you probably snacked more than you ate, and your parents may have commented on your small appetite for family meals and your larger appetite for sweets and other snack foods. You were the kid who wasn't interested in lunch, but was ravenous at two p.m. If so, you probably carried this style into adulthood, skipping some meals entirely, or thinking that a handful of nuts is sufficient at lunchtime, and then wondering why you have a headache and reach for a cup of coffee and vending machine food in the middle of the afternoon. It's not uncommon for people with a variable metabolism to be ravenously hungry at one meal and not interested in food at all for the rest of the day. Sometimes you feel hungry, but are so engrossed in what you are doing that you don't stop to eat. When you do eat, you are also multitasking, which means you are eating while standing at the kitchen counter, while sitting at your desk, or while on the phone. Throughout the day, you have enormous bursts of energy in which you make plans and burrow into work to the exclusion of everything else. When the energy vanishes and you crash, you reach for food to power you through the rest of the day.

Early in adulthood, people with a variable metabolism can get by even though they eat erratically. Because their young bodies don't readily store fat, they can have a horrible diet and not gain much weight. Sure, they might fantasize about losing five to ten pounds, but they've never been in danger of obesity. They are so busy talking, fidgeting, rushing around, and multitasking that they burn most of what they do eat. People with a variable metabolism often have very little experience with dieting as young adults. If you have a variable metabolism, you can cut out a couple of bad habits for a few weeks, or add a little exercise, knowing that your weight will normalize somewhat. That's just as well, because you may not have the patience for all the rules and food prep of a formal diet. Many people with a variable metabolism have gained and lost the same ten pounds many times. They love to start a new diet and exercise routine, but it never lasts long. If this is you, you may have always thought that the dirty little secret about you is that you have no willpower when it comes to dieting. No. The real truth about you is that you are easily bored, and most diets are fundamentally dull.

All of this changes as adulthood wears on. As you pass age thirty and particularly after age forty, the weight begins to creep on more steadily. Skipping meals sometimes and overeating other times puts stress on your metabolism, and your body may become more insulin resistant. You keep juggling multiple projects and multiple ambitions while rushing around, and you may find that it's more difficult to get to sleep at night. That will also cause you to gain weight. And you may find that sticking to an exercise routine becomes more difficult with your busy schedule. All of this means, too, that your bowels will start acting up. You may have more trouble with constipation, gas, and bloating.

People with a variable metabolism can also put on weight because of a significant change in mood or personal circumstances. They have moved to a new country, lost a job or a relationship, taken a job they don't like, or entered into a phase of life in which they have to be a caretaker to someone who is ailing. Stress exerts a powerful force on the body that further slows down the metabolism.

If you have a variable metabolism, here are some things you need to do to keep your body on track:

Set regular mealtimes. You have to give up on skipping meals, even if you don't feel hungry at first at certain times of the day. Set an alarm for yourself, if you need to. The idea that you can skip eating and save those calories for later in the day isn't working. You need to train your body to expect food at certain times. If you eat a light breakfast with good fiber, such as some oatmeal or a smoothie with lots of greens and fruit, followed by a substantial lunch, you won't be that hungry at dinner. So you can eat a little bit of soup or a light salad and your body will be satisfied. What's more, you will have more sustained energy in the heart of the day when you want to burrow into your projects and plans. Within a few weeks, you will begin to feel hungry at these times and be less tempted by snacky filler foods.

Prioritize fiber. People who tend to graze throughout the day rarely get enough vegetables and fiber in their diets. It's too easy to eat refined sugars and salty, flour-based snacks and call that a meal. This is why you may suffer from constipation. In addition, you may reach for heavy, grounding foods (pasta, pizza, muffins) when you feel scattered and overwhelmed. These things slow down your metabolism and steal your energy. Build every meal around vegetables, fruits, and whole grains. If you eat meat, make sure you do it in the middle of the day, when your body can best digest it.

Avoid dehydration. If your energy level varies all day long, you are battling dehydration, whether you know it or not. You will want to increase the amount of water you drink. Divide your weight in half, and that's how many ounces of water you should be taking in each day. Keep a water bottle near you while you work and you will find your mood and your thinking improved. Drinking more water will aid digestion and weight loss. Most people with this metabolism also need to reduce the number of dehydrating drinks they consume every day. That means less caffeine, alcohol, or soda and more herbal teas or hot water with lemon. Also, you have to rethink your relationship to dried foods and dried snacks. Avoid salty foods like chips, pretzels, and popcorn because they dry out your system.

Don't worry so much about fats. If your metabolism has its fits and starts, you probably don't need to restrict fats in order to lose weight or keep it off. You are lucky. In fact, most of my patients in this category who have been on a low-fat diet found that their skin became dry and their moods fluctuated quite a bit. Unlike people with a slower metabolism, you can eat some meat. Stay away from large portions of meat because your digestive fire isn't strong enough to burn it off. You can have some dairy for the same reason, but you should avoid ice cream because frozen foods weaken your digestion. You can have plenty of nuts and healthy oils without worrying about weight gain. Vata types struggle with dehydration, and your system is like blotting paper, absorbing those oils that will nourish your joints and tissues. You don't need fat-free salad dressing. Instead, enjoy the actual fats that feed your skin, your brain, and your joints.

Check in with your body before and after eating. This is good advice for anyone, but people with a variable metabolism are notorious for eating or grazing mindlessly. They eat at their desks while working; they eat standing up or rushing around. Instead, you need to sit down and really focus on enjoying the food in front of you. Always, always leave your place of work in order to eat. You can bring a lunch for yourself and eat it outside, or you can go to a restaurant to get a bowl of soup. Before you take the first bite, sit quietly and think about how your body feels. It takes just a few seconds. Make the connection between feelings and food. This is also useful on those occasions when you eat something you know isn't as healthy as you would like. Check in with your body before you eat it, then enjoy it, then check in with your body afterward. How do you feel? Since there is a difference between what feels good on the tongue and what feels good once it's inside the body, you have to make the connection between the two feelings in order to make better choices.

Strong Metabolism

If you answered B to most of the questions, you have what I call a strong metabolism. If so, you are a pitta, and "pitta" means fire.

Your body is a calorie-burning machine and always has been. Your body temperature runs hot, and so do your emotions. You are someone who can eat a lot, and who needs to eat in order to maintain your energy and focus. In fact, if you skip a meal, you may become irritable and unable to concentrate.

As a child, you were never a picky eater. You may have had favorite foods, and less favorite foods, but you were able to eat and digest almost anything that was put in front of you. You may have had phases in childhood where you became a little chubby, but then a growth spurt would stretch you out. As a teenager, you were the one who finished eating before everyone else, went back for seconds, and then, an hour later, were wondering when the next meal would be. You may have sneaked into the kitchen after hours to make a sandwich or raid the pantry before bed because the hunger just wouldn't go away. You may have relied heavily on sports drinks, sodas, and high-fat desserts to get the calories you needed for your growing body. You probably had a medium-sized, muscular build, or you might have been tall even as a teenager. You probably describe yourself as having a cast-iron stomach with regular bowel movements. Your body is digesting as fast as it can. You have always been active, interested in movement, achievement, and perfectionism. You may have been involved with sports teams or endurance events, and you are naturally competitive. All of that activity made you even hungrier. A lot of the people I see in this category learned to stuff themselves with food when they were teenagers and young adults. They learned to fear this feeling of being ravenous and to associate the feeling of being too full as one of satisfaction and safety.

As a young adult, you may have continued to eat large meals

with frequent snacks that included sports drinks, energy bars, and protein shakes. And you may have used exercise as your primary mechanism for controlling your weight. You may have told yourself that you can eat whatever you want, as long as you balance it with exercise, which was probably true for you as a young adult.

People who have a strong metabolism do experiment with dieting. Pitta types are natural perfectionists, and they are drawn to the idea of the perfect body or healthy food. They may be motivated solely by the number on the scale, and for a while they can deprive themselves and endure the hunger and irritation in order to achieve that one goal. People in this category are also drawn to endurance sports because an organized training schedule with a specific goal allows them to eat whatever they want without worrying about weight gain. They can do this for a while, and then other goals and stressors take over.

People with a strong digestion also love food. They love elaborate restaurant meals and create food rituals. They associate eating with pleasure and frequently don't think about the fact that food offers two pleasures, one on the tongue and another inside the body.

Balancing your strong need for calories and a desire for exercise works well until middle age, when the body's metabolism begins to slow down. At the same time, sports injuries become more common. You may have developed an overuse injury that keeps you from the gym or curtails the intensity of your workouts. This is when weight begins to creep on. There comes a point when intense exercise will not turn back the clock on your metabolism. By this point, many of the people I see have put on thirty or more pounds over the course of several years.

This is what had happened with Bruce. He told himself that as long as he was eating healthy food, he could eat as much as he wanted. Like many of the people of this body type, he had learned to eat beyond the point when he felt full. He would eat until he felt stuffed. In a way, that stuffed feeling was a measure of his success. If you love food, you need to move this love of food out of the restaurant scene and into your home. Cooking meals at home gives you the pleasure of being surrounded by food without eating a multicourse meal. You will also learn how to tailor portions to more appropriate sizes.

If you have this body type, you need these specific strategies to promote weight loss while preserving your strong digestive fire:

Eat clean foods. If your body is a furnace for calories and nutrition, you must feed it clean fuel. That means cutting down on red meats, pork, and shellfish. And it means turning away from heavy sauces and fried foods and most junk foods. You want to be focusing on proteins in the form of white meats, egg whites, and tofu. Eat cheeses with a low level of salt, so that you aren't retaining lots of sodium. Instead, you should look to build meals around fruits and vegetables, which will allow you to feel more full without clouding your body with sauces and heavy fats that are making you feel dull and probably interfering with your sleep.

Plan to eat your largest meal at noon. Yes, this is important for everyone, but it is nonnegotiable for people with a strong metabolism. If you are used to saving your largest meal for the end of the day, you are working against your naturally high metabolism. Taking in lots of calories after work is keeping you from losing excess weight. It's also contributing to indigestion at night, heartburn, and congestion. Simply by moving your largest meal earlier in the day, you can jump-start any weight loss you hope to achieve.

Learn to tell the difference between real hunger and fake hunger. If you've spent years eating large meals with frequent snacks, then your body is used to having high blood sugar all day. Whenever your blood sugar levels drop a bit, your body panics and tells you that you are starving, even though you aren't. Don't worry, your body can learn to tell the difference between real hunger and the fake kind. As you train your body to follow a stricter schedule, I recommend you keep a thermos of green tea sweetened with honey on hand. You can sip this all morning instead of coffee and it will keep you away from the vending machines. In the afternoon, switch to drinking water with lemon, which will help regulate your blood sugar and keep you from feeling hungry between meals. Usually a few sips of water will take fake hunger away.

Cool your digestion. Your metabolism runs hot, like your emotions. So you want to stay away from spicy foods that can cause you to flush and sweat. These can also cause congestion and heartburn. Your body wants milder, savory spices. When you are looking to clear your diet of unwanted snacks, focus on raw foods and reach for fruit and raw vegetables to help cool you down and settle your hunger. They last longer in your system than salty snacks and sweets.

Eat sweets with complex tastes. If you have a sweet tooth, you may reach for desserts or sweet treats throughout the day. Many people with a strong metabolism have a sweet tooth. But it's important to remember that there are all kinds of sugars. You don't have to avoid all sugars. It's important to choose the right kind. Refined sugars and refined flours will make your sweet tooth stronger. Fruits, by contrast, contain lots of complex tastes. An apple contains sweetness, but is also acidic, crunchy, and sour. When you choose foods for a meal or for a treat, choose those that have lots of complex tastes and textures. These will help you feel satisfied.

Slow Metabolism

If you answered C to most of the questions, you probably have a slower metabolism. In Ayurveda, you would be called a "kapha," which means water. If you have this constitution, your body more readily stores both fluids and fats. In terms of build, you may have a body that is stocky, with larger bones and joints. You might have large eyes, thick hair, and smooth skin that is cool to the touch. Your metabolism also allows you to work harder and longer on a project than anyone else. You can handle stress and discord better than anyone in a workplace. You are not at the mercy of your moods, like other body types, and you can actually remain healthier longer.

As a baby, you would have been a joy. You were a good sleeper and generally content. You were an easygoing toddler who was curious and happy. Once you learned something, you could remember it forever. You were never a picky eater and grew strong quickly. As a child and teenager, your relationship with food may have become more fraught. In a culture that prioritizes thin bodies, and even skinny bodies, you may have struggled at a young age with the idea that your body should look different than it does naturally. With your larger bones and joints, it would be impossible to have a very skinny, small frame. And yet your parents may have encouraged the idea that you should be thin and that you should eat less than you do. This has been overwhelmingly true for women, but even young boys are being taught that they should have thin waists and visible muscles. As a child, this wasn't really how your body looked. If you had siblings or close friends with different body types, you may have wondered why they could eat more than you did and yet remain thin. Your passion for foods and sweets was probably no greater than theirs. People with a slower metabolism may struggle particularly in the first twenty years of life with this notion, because this phase of life is when the body naturally stores fats and fluids in order to grow.

Because your body retains fluids, you may tend to get colds in the winter. You have a respiratory system that produces more mucus and is sensitive to pollen and toxins in the air. You are more likely to gain weight in the springtime, when the weather is damp and humid; and if you move to a climate that is frequently damp and humid, you may experience even more weight gain.

As a young adult, you may have done a lot of dieting. The people I see with this body type have tried all the fad diets and have flirted with extreme exercise once or twice to try to get their bodies to look the way they feel they ought to. The trouble with extreme diets and so-called diet foods, such as meal replacement drinks and bars, is that they contain synthetic nutrients, and low-calorie foods contain artificial sweeteners. Nothing in these foods satisfies the body or the palate. And they can actually trigger binges later on. That's why binge dieters so often yo-yo in weight. You may have starved yourself without finding satisfaction in whole foods and then binged when you could no longer hold out.

On the other extreme, I see people who have given up on dieting. They are understandably frustrated because when they do diet, they have to be absolutely vigilant in order to lose weight. At the same time, the people around them seem to be able to indulge in sweets and unhealthy foods almost at will. What's worse is that they feel judged by others who eat more than they do, and who can make terrible food choices without suffering the consequences.

If you have given up on dieting and exercise, and you have a slow metabolism, you can gain five to ten pounds every year by eating the way your friends and family members eat. By doing this, you can easily drift into obesity well before middle age. What I tell people who struggle with a slow metabolism is that there are ways to diet that give you the best chance of successfully maintaining a healthy weight. In order to do all of this, you need a diet that removes toxins and discourages your body from retaining water weight.

Kapha types will do best if they:

Detox first. Because you have a body that retains both fluids and fat cells, you are retaining the toxins you have ingested from unhealthy foods. In Ayurveda, we refer to toxins as *ama*, and you may notice ama in your body as congestion, a whitish or yellowish coating on the tongue in the morning, or just in a bloated feeling or heaviness in the morning or after eating. In fact, if you are building up ama, you might notice that you weigh more in the morning than you did the night before. You need to flush this from your system with a detox diet that bans refined sugars and flours, most meats, and most dairy. It sounds severe, but when people with a slow metabolism give up flour and dairy, their bodies respond almost immediately. Dairy products are notorious for causing mucus to form in the body, and most people in this category do begin to notice that their digestive tracts react to dairy and meat products and they may even have heartburn or congestion shortly after eating them. Many of my patients find that they do quite well on a vegetarian or vegan diet as long as they also de-emphasize pastas, baked goods with wheat flour, and rice. If you need a jump start, you can use the detox diet outlined on page 149. You may want to adopt this diet for three to four weeks and see how your body feels. People are often surprised at how well they do with it.

Ask yourself if you are really hungry. While this is good advice for everyone, it's crucial for kapha types. I've found that people with this body type aren't as hungry as they think they are. They have trained their bodies to eat three times a day, or more, but when they connect to their bodies, they find that they don't have that much of an appetite. They tend to graze through a meal rather than gobbling food down the way others do. If you have this metabolism, you may have told yourself that you don't have willpower or that diets have to be miserable, but that's not necessarily true for you. Your body contains a great deal of wisdom about what it needs, and if you check in with yourself before and after each meal, asking, *How do I feel?* you will be surprised at what you learn. Your body can thrive with a lot less food than what is prescribed by the typical Western diet. When you eat cleanly and only when you are truly hungry, your body will reward you with renewed energy and clearer thinking. While it may take a couple of weeks of clean eating to retrain your taste buds, you can do it. I've seen it many times.

Embrace gentle fasting. The real power of this body type is the ability to go without food from time to time without suffering. If you miss the evening meal, you won't crash like people with a variable metabolism, and you won't become irritated and moody like people with a strong metabolism. You will continue to be the same gentle, focused person you always present to the world. Other people can't skip meals because their brains and bodies become fogged, but you can. In fact, many of my patients try skipping dinner a couple of nights a week and they actually feel better. They feel lighter, as though their bodies and minds function better than ever.

Prioritize bitter and astringent foods. Other body types should probably stay away from coffee, but if you have a slower metabolism, you're in luck: black coffee is good for you. It's a natural diuretic, which means that it flushes fluids from your system that your body would otherwise retain. You can also drink a lot of black teas and green teas in the morning to get the metabolism fired up and to release fluids from your system. In terms of fruits, you can have anything astringent, such as apples, cranberries, grapefruit, and most dried fruits in moderation. You want to gravitate away from sweetness toward other flavors. Hot water with lemon and ginger can be a mainstay of your daily regimen. Drinking this will help you retrain your taste buds.

Enjoy spices. If you love big flavors in your food, then enjoy as many spices as you like. This body type does well with any type of food that is hot and spicy. Hot spices cause mucus to flush from the system. I tell people to eat whatever creates a warming or flushing sensation in the body. This means vegetarian chilis, or any hot liquids. I tell people with this body type that Thai food is great for them. Get the hot vegetable curry without rice because it is spicy and satisfying. You want to be eating those foods that open up the airways and improve your breathing.

The Detox Diet

Most of this book is focused on aligning your body with its circadian rhythm so you can sleep better and lose weight naturally. But some of the people I see in my practice do need to recover from years or even decades of poor eating. If you need to lose more than thirty pounds, it's possible that your body needs more than a critical realignment of the circadian rhythm. Before you are ready to eat right for your body type, you may need to retrain your body to better digest food and help your taste buds to better appreciate good food.

When people come to me with serious weight-loss issues, I put them on a detox diet for several weeks. It's an eating plan that resembles the popular low-carb diets or what some people call "slow-carb" diets. By eliminating all the foods that trap toxins in the body, you can not only jump-start your weight loss, but you will sleep better and have more energy to exercise. Remember, though, that this is a temporary diet. In fact, not everyone needs this diet. It works best for people who have gained so much weight that they are unable to fully digest the food they eat every day and therefore can't see results from typical diets.

If you have a variable metabolism, you should attempt this diet for no more than one to two weeks. If you have a strong metabolism, you can stay on this diet for up to four weeks. If you have a slow metabolism, you can stay on this detox diet for as long as you like, but you may want to cycle off of it from time to time and then restart it when the seasons change. When the body is moving from one season to another, you may be more likely to build up toxins, so that's a good time to start a detox like this.

In Ayurveda, we talk about emphasizing intelligent foods, those foods that contain phytonutrients because they are alive and vibrant. No matter who you are, you want to choose foods that are grown in nature and unrefined, but this is especially true while trying to detoxify the body. The trouble with eating a diet high in processed and refined foods is that your body can't readily digest these ingredients. They leave residue in the digestive tract. Whole foods, by contrast, can be fully digested and leave the body cleanly in waste. I like to tell patients that the digestive tract is like a carpet, and the residue from bad foods gets stuck in the carpet of your body. The added bonus is that it's difficult to overeat on this diet.

With this in mind, your detox diet should emphasize foods found in nature and those most easily digested.

Vegetables: You can't have too many of these. You will be looking to eat them raw or steamed lightly. And this will be the staple for every lunch and dinner. If you need to dress them, use something vinegar-based with very little oil. Sprouts count as vegetables and are very healthy and detoxifying. You can have vegetable juices as well, but stay away from carrot juice. The only vegetables you need to avoid are heavy, starchy vegetables, such as potatoes and sweet potatoes.

Spices: You can use spices liberally in this diet. Use them to enhance the flavor of the vegetables you eat. The one exception is salt, which you want to use in small amounts because it can cause bloating.

Whole grains: Most people understand that refined flours are not good for people who want to detoxify their systems and lose weight. So you want to avoid breads, pastries, and pastas. You can eat steel cut oats sparingly, and people in this detox phase should avoid all flour-based products entirely. You can eat many whole grains, including barley or basmati rice mixed in equal parts with mung beans and cooked into a kind of lentil soup (called *kitchari*). Basmati rice is an excellent detox grain for this phase because it is so easy to digest.

Nuts: Raw nuts, such as sunflower seeds, pumpkin seeds, and sesame seeds, may be eaten occasionally. They work well as a topping sprinkled lightly on oatmeal or vegetables.

While you are detoxing, you want to avoid certain foods despite the fact that your taste buds love them, and even some foods that you may have thought were healthy, including:

Flour: Most people don't know how much they rely on flour-based products as a diet staple. Corn chips, pita bread, breakfast cereals, even whole-grain breads, contain lots of refined grains that sabotage any weight-loss or detox goals. They spike your blood sugar levels and increase your hunger later on. These things fill you up but don't provide nutrition.

Dairy: Milk and other dairy products produce mucus in the system, particularly when they are pasteurized. Yogurt is often considered a health food, but the kinds of yogurt most people buy contain too much added sugar and very little healthy cultures. As such, it's no better than any other kind of dairy. It's better to avoid yogurt altogether while you detox. It has a reputation as a health food because when it's freshly made at home, it contains lots of live bacteria that are good for your digestive tract. Everyone can use a little more of these healthy bacterial cultures, but you can get these from kimchi or kombucha. You can even take acidophilus pills. While almost all dairy is off the menu, you can have unsalted Ayurvedic buttermilk or ghee in small quantities.

Meats: Animal products often cause the body to build up toxins, so for this detox phase, you should avoid all meats, including fish and shellfish. If you feel you need more protein (for example, if you're an athlete), chicken and turkey can be eaten in small quantities, but in general you should avoid all meats.

Fruits: These are normally very good for you, but in the detox phase of your diet, you want to use them only as desserts or treats. You especially want to avoid heavy fruits such as bananas, avocados, and most fruit juices. If you do eat fruits, you want to gravitate toward the astringent fruits, such as apples, citrus, cranberries, and pomegranates.

Beans: Although beans contain a tremendous amount of fiber, they can cause gas and bloating. I suggest avoiding beans as well in the detox phase, except for mung beans. Lentils and mung beans are smaller and more easily digested; therefore they cause less gas and bloating than larger beans. Many of my patients trying to detox find that a lentil soup with plenty of vegetables becomes a staple for them. Eventually, it becomes a kind of comfort food.

Most fats: Unhealthy fats are difficult for the body to digest and they also cause mucus to form in the system. Of course, this means fried foods, but it also means butter, margarine, cheeses, and animal fats. You can have raw, cold, unprocessed oils in small quantities. This includes ghee, or clarified butter, which is a healthy choice.

Sugar: White sugar is one of the most toxic foods, and should be avoided in this diet. If you need a sweetener, you can use honey instead, but don't heat it or cook with it. Honey is a delicate food, pulled from the essence of many different plants. Despite being sticky, heavy, and sweet, honey doesn't inflame the body or cause mucus. Rather remarkably, it has a drying effect on the body when consumed at room temperature. However, all of this changes when you cook with it. Heat degrades its antioxidant power along with its flavor. So you don't want to use it in baking or cooking.

Cold drinks: You want to increase the digestive fire, and cool drinks have the opposite effect on the body. They chill the digestive acids in the stomach and shock the system. It's better to drink liquid at room temperature or heated. Hot water with lemon will help cleanse your system as you detox. Avoid coffee; you can drink green tea instead if you need caffeine.

Instead of trying a one-size-fits-all diet, you can find the right diet principles for your body. When you add these healthy eating habits to an eating and sleeping schedule that fits the circadian rhythm, you'll find that maintaining a healthy weight is easier than it has ever been. Once you've given up late nights in front of the TV and evening snacking, and you've replaced these habits with a small, nutritious breakfast, a larger, healthy lunch, and a small dinner, you will find incredible resources of energy and focus in your life.

The final factor in your new lifestyle is exercise, which I'll describe in the next chapter. I'm not talking about long hours at the gym. Instead, I'll outline how a reasonable daily exercise regimen will build on all that you've accomplished so far. After all, exercise is a primary means for detoxifying the body. It also promotes healthy eating and deep, restorative sleep.

8
The Right Exercise at the Right Time

I'd be willing to bet that you know that exercise is good for you, even if you sometimes avoid it. What you probably don't know is that in Ayurveda, exercise is considered a sacred daily ritual, and one of the most powerful ways to keep your body functioning as it should. The Sanskrit term for this is *vyayama*, and it would take many paragraphs to describe its nuances. Put simply, it means improving circulation and communication through specific movements. We all know that exercise improves the circulatory system, getting your heart pumping and your lungs working. Yet that's not the only system in your body that needs you to be active. The vedas speak about each system in the body as a channel. The digestive system is a hollow channel through which nutrients enter and pass through your body. The respiratory system is another channel for oxygen and carbon dioxide. The lymph system, the nervous system, and the circulatory system are all channels through which the body passes nutrients, fluids, signals, and waste. Your body has been designed to do all of these amazing things, but it needs a little help.

Over the course of a single day, these channels get sticky, or slimy, or blocked. This blockage is a natural by-product of daily life, but, just as it's important to brush your teeth to prevent tartar buildup, it's important to clear these blockages daily. You do this using prana, which means life force, but is sometimes translated as "breath." And for good reason! Through exercise and movement, you breathe deeply and unblock the body's channels. Exercise is essential because it causes deep breathing and triggers important chemical and hormonal changes that your body needs to stay healthy. Through exercise, you improve communication between the systems so they can operate in harmony. Exercise also warms the body, creating agni—or "fire"—and therefore stokes your digestive fire. It gives you energy, mental clarity, passion, and a general enthusiasm for life. While it's tempting to think of exercise as something to cross off of your to-do list, it actually needs to be integrated into your daily schedule. It offers an emotional reset. It improves energy and mood. It cleanses the body.

Exercise becomes even more important in a modern schedule that leaves little time for movement. Commuters sit in their cars or on trains to get to work, then sit at a desk all day, then at night they sit in front of a television or computer screen. It's hard to get restful sleep if you've done nothing to tire the body's muscles, and it's impossible for the digestive system to move effectively without physical activity to help it along. Ayurveda is the art of bringing balance to your body. If you are cold, you must bring in heat to create balance. If you are dehydrated, you need fluids and oil to bring back balance. If you are sitting all day and sleeping all night, you require movement to restore balance. And the best time for this first intense bout of movement is in the morning, after you have been sleeping for seven or eight hours.

Prana Builds Energy

In Ayurveda, the two most important organs in the body are the heart and the brain. Only prana, the breath, connects them both. When you breathe deeply, you connect your heart and your brain instantly. You can do this right now. Take a slow, deep breath. Enjoy it. Take another one if you like. You can probably feel the stress melt away a little bit. Slow, deep breaths are the cornerstone of meditation and yoga, but they are instantly available through even the most modest level of exercise. Conscious exercise puts your daily problems on the back burner while it enlivens the heart and mind. Is there anything more immediately healthy than this? By contrast, a sedentary life causes you to breathe only shallowly, and this strains the heart and starves the brain.

But prana means more than connecting to your breath or taking deep breaths. The Sanskrit word also refers to the vital energy that sustains life. Without daily movement, you are failing to take these deep breaths and you are draining your energy. Because you are disconnected from your body, you don't even notice it. You think you are fatigued or bored at the end of a day of sitting, but it's really more than that. Your brain is starved of oxygen, and so are the tissues in your body. They get blocked and unable to function. Low energy causes you to reach for multiple cups of coffee throughout the day to jump-start your brain. This isn't just an Ayurvedic view of energy. New research shows that even low-intensity exercise can reverse symptoms of general fatigue if you do it every day.[1] Even a short daily morning walk will get your breath going, get prana into your body. It will help you sleep at night, help you to sync with your body's daily rhythm. It will make you hungry for breakfast in the morning and give you focus for the day.

Instead of doing a short burst of daily exercise, many people try to inject two or three intense exercise sessions into their week, when exercise is convenient. Guidelines stress how many minutes per week you should get of exercise. If you make two to three trips to the gym every week, you may wonder why this doesn't change your waistline or daily energy levels. On the contrary, longer sessions of intense exercise can wring you out, making you exhausted. You will be less energized and more likely to skip going to the gym on stressful days when you need it most.

One of my patients is a wealth manager who travels across the country twice a month on business. He is on a plane moving from one office to another every other week. He knows that he eats too much and sleeps too little and, in an effort to combat low energy levels and his expanding waistline, he took up strenuous exercise. He bought an exercise bike and two or three times a week, he logged into an online class and the instructor in another city led him through an intense one-hour workout. Although this intense exercise helped him take the edge off of his frustration and worked up a nice sweat, it didn't help him lose weight. In fact, it didn't cure his insomnia, his headaches, or his struggles with sexual function, either. He didn't want to take any little blue pills, so he came to see me. While there are a number of problems with his schedule, including chronic jet lag, late-night eating, and late-night computer work, I zeroed in on his exercise schedule first, because all of his problems centered around having no energy. If you are overcommitted and exhausted, you will have insomnia, low energy, a short temper, and headaches. When your fatigue becomes extreme, you will have problems with sexual function. You need more circulation in your body on a daily basis. You need to breathe deeply and clear the body's channels.

The counselor in me wanted to focus on getting him in touch with his body and how he actually feels while eating and working, but that would involve asking him to meditate or take his pulse, and I knew he wouldn't do that. What he could do, though, was make morning exercise a staple in his schedule. Not surprisingly, when I suggested this driven wealth manager try a morning yoga class where he could breathe deeply, stretch his muscles, and feel the body he was ignoring the other twenty-three hours of the day, he balked. The entire fitness industry was shouting at him that if he didn't push himself to extremes with each workout, he wasn't getting "real" exercise.

No overachiever likes to be told to do less, but many of my type A patients need to heed this advice: exercise is meant to be stimulating. Though it may seem counterintuitive, less intense exercise sessions every day will improve your energy level, and it's crucial to be doing it earlier in the day so that it can help you sleep at night. If you aren't doing any exercise, you already know what the problem is. It's harder to know if you are doing too much, or doing the wrong type of exercise. If you are not gaining all of the physical *and* emotional benefits from the exercise you are doing, then it's not working for you. If your exercise routine doesn't help you sleep at night, if it's not changing your mood for the better or giving you more energy, then you need to adjust something. The wealth manager did end up going to a yoga class twice a week and absolutely loved the deep breathing. I told him it was like a meditation with his body, and he liked that idea because he said he got bored with the traditional sitting meditation. He was so used to this multitasking mentality that he needed exercise that's also meditation. On the rest of the days of the week, he does a short walk or some jogging. And his energy has skyrocketed. While it's tempting to think of exercise solely as a means to fitness or weight loss, it's really much more about making you feel alive in your body.

The Circadian Connection

I've described how sleep (or lack of sleep) affects the body's circadian rhythm. So does eating or restricting the hours in which you eat. A third mechanism by which the body tells time is activity. When you are physically active, the body logically assumes that it's daytime. This is one reason why I urge people to get out of bed and do some exercise before they eat, before they do anything else. It gives the body an undeniable signal that the day has started.

This morning exercise will also help you get to sleep at night. In some studies, mice who have the chance to engage in a period of exercise go to sleep earlier and get up earlier than those who don't exercise. Even a single bout of intense exercise can shift a mouse's circadian rhythm. Human studies aren't quite this dramatic, but it's clear that regular exercisers report less trouble getting to sleep. In one study, people went to live in an isolation facility so they would have no idea what time of day it was. They lived on a slightly shortened day, so that each night they were told to try and get to sleep about twenty minutes earlier than the previous night. Think of it as experiencing jet lag in short, daily intervals. Some were told to exercise and others were not. After about six days, those who were required to exercise were adapting to the changing sleep cycle much faster than those who did no exercise.[2] The change was happening inside their brains with a higher production of melatonin (the brain's natural sleeping pill). Their brains were releasing melatonin earlier, which made them able to sleep earlier.

Here's where it gets really interesting, though: these same subjects were then asked to exercise at different times of the day. Those who exercised in the afternoon or at night did not get the same shift in melatonin production. In fact, intense exercise close to bedtime actually delayed the production of melatonin, making it harder for them to go to sleep when the lights went out.[3] It turns out that exercise at the wrong time of day confuses the body's central regulator of the circadian clock, which in turn confuses the body's cells and systems.

This is important to know if you like to hit the gym after work. Many people think of exercise as something to squeeze in at the end of the day. Anyone who has ever fought for a treadmill after work knows that the busiest hours at the gym are between six and eight p.m. For a large chunk of the year, this means you are exercising after dark, at a time when the body is winding down and preparing for sleep. This intense activity may make you too alert and overheated for sleep.

Recently, I met with a woman who seemed to be doing everything right. She was in her early thirties, an avid exerciser, and had a diet that anyone would envy. She was a lifelong exerciser, someone who said she needs strenuous workouts to help her deal with stress. Her one problem was that she was not sleeping at night, despite a careful and consistent bedtime routine.

When I looked more closely at her schedule, I found that she was leaving for the gym at eight p.m., which meant that for most of the year, she was exercising during the body's rest phase. By the time she got home, her body was wide awake, still sweaty, and hungry on top of that. She knew she shouldn't eat again after her workout and would try to go directly to bed. Given her body's structure, it was true that she was getting a lot out of her strenuous exercising, yet it was giving her insomnia in the short term and making her foggy and lethargic the next day. The insomnia was new for her, and it's possible that she was able to stick to this exercise routine in her twenties without noticing much difficulty with sleeping, but the body is changing in your thirties, when it starts to be more sensitive to sleep disturbances. By getting up earlier and doing her intensive exercise first thing, she primed herself to be more awake at work, but less awake and less hungry at the end of the day.

Morning exercise is especially beneficial to your circadian rhythm as you age. In your late forties and early fifties, you will need to give your body strong signals in the morning that the day has started so you can get to sleep at night. This will counteract the hormonal changes in midlife that can cause insomnia. A recent study looked at the effect of exercise on reducing cancer risk in older women. Fortunately, researchers included questions about how participants slept during the study. That's how we know that postmenopausal women who did morning exercise had fewer complaints about insomnia. These were women between the ages of fifty and seventy-five. Those who did forty-five minutes of moderate exercise in the morning five days a week slept better at night. Even those who restricted themselves to light stretching in the morning slept a little better at night.

Those who exercised in the evening had more trouble with insomnia.[4] When you get up and get going in the morning, you'll have an easier time winding down at bedtime.

Clock Genes and Metabolism

The most recent research goes inside the cells of skeletal muscles to find out what they are doing during intense exercise. When you are doing low-intensity exercise, such as a leisurely walk around the block, your muscles use oxygen as fuel. If you switch from walking to jogging or sprinting, you will start to gasp for air because your muscles are depleting their oxygen stores. When this happens, your muscles switch to using available sugar as fuel to keep working. Researchers who are looking at these metabolic changes have found that skeletal muscles contain clock genes, just like all of your other cells. Actually, the clock genes are more involved in metabolism than you think. It's these clock genes that signal to the cell to switch from burning oxygen to burning sugar during exercise.[5]

Remember that your clock genes are signaling the cells to do one thing during daylight hours and something else at night. So your muscles have a differing ability to adapt to exercise and to use energy throughout a twenty-four-hour day, just like every other system in the body. Muscle cells are more efficient during daylight hours, which means that they are better able to contract and to make this metabolic shift from oxygen to sugar. That's why intense exercise is more beneficial to your metabolism as a whole when it's done earlier in the day, and it can help control blood sugar levels. In fact, this research suggests why a total lack of exercise has such dire metabolic consequences. Exercise is such a reliable means for controlling blood sugar levels that we may come to realize that a sedentary life is the primary risk factor for developing type 2 diabetes.

But there could be an even greater benefit to exercising first thing in the morning. In chapter two, I talked about a study that showed how men could reverse the effects of a high-fat diet by exercising before breakfast. Rather miraculously, they didn't gain weight if they exercised before their first meal, but if they exercised later in the day, after eating one or two meals, they did gain weight. This new research on clock genes and how they affect metabolism may hold a clue. When you exercise before breakfast, you are exercising while the body is still fasting. So when you exercise hard enough to trigger your muscles to switch from burning oxygen to burning sugar, your blood has less available sugar to feed them. As a result, the body is forced to tap into fat stores.

The good news is that you don't need to do a lot of intense exercise before breakfast to get this result. Just twenty to twenty-five minutes will do the trick. You could be on a treadmill or outside. If you are relatively new to exercise, you can do interval training, in which you are walking as fast as you can for a minute and then walking at a more leisurely pace for two minutes. Or you could be jogging and alternate that with walking. You can work on a rowing machine with greater and lesser intensity. You don't need to kill yourself. The goal is to sustain a burst of intense movement until your breathing changes. At the point when you feel that you need to breathe out of your mouth, you should slow down for the recovery period of lower activity. Panting and straining on a treadmill isn't truly exercising—it just puts undue stress on the body. You want to be working at the edge of your ability, working in bursts and then recovering. When you do this, you'll have about seven minutes of working intensely in your twenty-minute workout. That's enough to make the metabolic changes you need, and it's enough to improve your fitness day by day.

What we are looking for is the minimum effective dose of exercise for you and for your body. I'll talk more about the right exercise for your body type in the next chapter—as you know by now, not every body needs exactly the same kind of exercise—but there are some universal tips to help you sync your circadian rhythm and trigger those metabolic changes that will help you regulate your weight. Many of my patients learn to love their early morning exercise because it's easier than any workout they've ever tried before, and because it wakes them up better than a cup of coffee. The point is to get up and do something that opens the channels, that causes you to breathe deeply and to sweat a little bit.

How Much Daily Exercise Do I Need?

This is a tough question to answer because many people think of exercise as a formal activity. You change into workout clothes and go to the gym or to a class, after which you return home, shower, and change again. But in Ayurveda, exercise is what you do whenever you move your body. You might be sweeping the floor, mowing the lawn, taking a stroll after dinner, or playing catch in the backyard. All of these things move the body and open its channels. And you can do these things at any time of day. If you've ever purchased a fitness tracker and become suddenly obsessed with your daily count of footsteps, you are taking the Ayurvedic view of movement. It all adds up. The body needs cycles of activity and rest, and that's one reason why a single intense bout of morning exercise is so important. It starts the cycle of activity and rest first thing in the morning. Your body will prompt you to get up and move throughout the day, so make sure you get up at regular intervals to take a short stroll, take the stairs, or do some light stretching. In total, you need an hour of this movement every day. Most people don't even think about how much they are moving their bodies relative to how much time they spend sitting.

A recent report compiled data from sixteen different studies about the effect of exercise on longevity with fascinating results. The one million participants were divided into groups based on the amount of exercise they did. Those who were active for sixty minutes per day completely reversed the effects of a sedentary lifestyle and lived far longer than those who did little or no exercise.[6] But what was interesting was that these top exercisers were not necessarily hitting the gym or doing intense exercise classes every day. In fact, they weren't always breaking much of a sweat. The moderate exercise included walking about three and a half miles per hour, a very doable pace. In some cases, it meant cycling at ten miles per hour. The benefit comes simply from moving your body. It's not clear from the study whether participants did all of this exercise at one time. In my experience, people who find a way to stay naturally active through walking, yoga, and other light exercise every day are in the best position to age gracefully and live longer.

If you are tracking for yourself how much movement you

get in a day, you should note what you do before breakfast, then track how many times you get up and move around during the day. Sitting at a desk all day makes you lethargic. You may have felt your attention wandering and your energy failing at midmorning and midafternoon. This is when many people rush to the coffeepot or the vending machine to prop up their flagging energy. But what your body is crying out for is movement. That's why you feel sleepy or achy. Take these emotional urges as the wake-up call they really are. Go up and down the stairs. Go walk around the block. You can do some simple exercises at your desk to get yourself to take those deep breaths your body needs. Remember that prana builds energy. You have the power to move your body and trigger the deep breathing that your body craves throughout the day.

But Wait, More Cardio Is Still Better, Right?

If you think of exercise only in terms of minutes on the treadmill, you may wonder whether you should do as much as you can in order to get more out of it. Researchers have wondered about this, too. If some is good, then more must be better. In fact, there is a popular myth that you have to do an hour of cardio a day in order to lose weight. But this was called into question by a recent study carried out with overweight but otherwise healthy men who didn't exercise. One group was asked to work out for thirty minutes a day, hard enough to break a sweat. The second group worked out intensely for a full sixty minutes every day. In the end, both groups lost weight, yet those with the shorter workouts lost slightly more weight than those who spent the full hour on the treadmill.[7] The researchers couldn't really account for these results. They expected that more intense exercise would provide more benefit, although they were savvy enough to realize that it probably wouldn't provide twice the benefit. Instead, they found that it was actually slightly less effective to expend twice as much energy and time in intense exercise. They speculated that perhaps intense exercisers ate more throughout the day because they'd burned so many calories through exercise. That's possible. But I know that newly motivated exercisers often push themselves past the point where exercise enlivens them. Instead, they work until they are exhausted and lethargic. Rather than having systems that are openly communicating, they have systems in crisis. Their bodies need to conserve energy for hours afterward. If you come home from the gym feeling wrung out and defeated, you may not be doing as much good as you hope.

The bottom line is that a shorter, daily burst of intensive exercise may be best, but you probably don't need as much as you think. Thirty minutes is probably enough as long as you augment it with more movement all day long. The trouble with most exercise routines is that they tend to be an all-or-nothing proposition. You decide to get in shape, and then join a gym and work hard until, gradually, you find conflicts in your schedule and skip a few workouts. Eventually, you forget to go at all. Weeks or months later, you repeat the process. Millions of gym memberships languish in this way. In fact, gym owners count on signing up hundreds of new members every January, knowing that these folks will come to work out for a few weeks or months and then never show up again.

According to most guidelines, people need between two and a half to four hours of exercise per week. If you break that down into a daily amount, you are looking at twenty to thirty-five minutes a day. Some people do all of their exercising on the weekends in an effort to reach these goals. They live the usual sedentary lifestyle during the week and then take multiple fitness classes on the weekends, coupled with running or distance biking. Then they wonder why exercise isn't changing their energy level or waistlines. Marathon exercise sessions don't make up for missed days. Instead, they put you in the path of overuse injuries and exhaustion.

It's hard for most people to start and stick to an exercise routine. The trick is to do some exercise every day, even if it's less than what you were doing before. If you haven't exercised in many years, start by walking briskly every morning, even if it's just for ten to fifteen minutes. If you take a more leisurely walk after lunch and another after dinner, you can move your body for thirty to forty-five minutes each day. By doing this, you've unblocked your system each morning. You've breathed deeply and cleared your mind. Some of my patients start their new exercise routines with a few yoga poses in the morning and that's it. My wife likes to walk on the beach in the morning. She tried to go to the gym every day, but it left her exhausted. A daily walk in a beautiful setting is what she needs most to become energized and centered for the day.

When you start a new exercise program, do what you can do every morning and think of that as your exercise. Gradually, you can increase intensity and reap even more benefits from strenuous work. You can add some isometric exercises to your stretching. You can jog a little bit. You can still go to your favorite fitness classes on the weekends, but you won't be thinking of it as making up for lost time. The key is consistency. Add to this a little bit of exercise throughout the day. Light housework or gardening counts as movement. I advised one couple to take a short walk after dinner each night instead of settling in to watch TV. It quickly became their favorite time of day, a time when they could reconnect with their bodies and with each other.

How Can I Make Time for Morning Exercise?

I hear this question nearly every day. People have so many good reasons why getting up a little earlier and exercising first thing is impossible for them. The kids need help to go off to school. The commute is lengthy. There are babies sleeping. It's too cold or rainy for large parts of the year. It's a hassle to put on the exercise gear and drive to the gym. And it's true that most people think of going to a gym or fitness class as a multihour event: changing clothes, driving, working out, and then reversing the whole process. But remember, prana-fostering, energizing exercise doesn't have to be such a production. Brisk walking in the morning on days when you can and finding some other way to get a short burst of intense exercise—such as jumping rope or a few isometric exercises at home—on other days, will do just as well. You may have to be creative about this at first, but the health benefits are tremendous, particularly for your body's central clock. You can get some sort of exercise machine, a little rower or stationary bike that can sit in the corner of one room. Hop on this for twenty minutes in the morning; you can even do it in your pajamas if you want. My patients who do this say it wakes them up better than coffee.

If you have multiple health-related problems, including weight gain, insomnia, low energy, and poor diet, you need exercise every day. Nothing invigorates your body faster than daily exercise. It will change your life. Some people like to say that meditation is the key to stress management and enlightened health choices. I agree. But if you are twenty pounds overweight and you sit down to meditate in the morning, you'll be snoring in two minutes. (That's why sun salutations or deep breathing are often advised as a prelude to meditation, so you can wake your body up before you clear your mind.) People say that diet choices are the key to losing weight. Again, I agree. But if your body feels dull and heavy, you will need herculean strength to eat better food all day. Exercise quickly aligns you with your sleeping and waking cycles and creates energy in the body, lifting your mood and opening the channels so that you can make better eating choices throughout the day. You will also have deeper and more meaningful meditation sessions if you have exercised first.

Aging and Exercise

Your body's response to exercise changes as you age. But that doesn't mean you should stop exercising after you turn fifty or sixty. Far from it. As you pass through middle age, your body's natural circadian rhythm begins to weaken, and that's when you need exercise to give it a boost.

As you age, you want to pay attention to your body's level of inflammation. Chronic aches and pains can be a sign of inflammation throughout your body. When active people come to me saying they have joint issues, I immediately wonder about their level of exercise. If they aren't doing any exercise, they are setting themselves up for age-related aches and pains. And if they are doing too much intense exercise, they are probably dealing with pain from systemic inflammation. It's a tricky balance, and I urge you to pay attention to how you feel while you are exercising and afterward, even if you enjoy your workouts.

A woman came to see me a couple of years ago because she was taking lots of over-the-counter pain relievers for sore knees. She was a lifelong tennis player, someone who still competed at a fairly high level at the age of sixty-one. She was practicing twice a week and competing once a week, which is amazing. Unfortunately, she had begun to take ibuprofen routinely, both before and after her matches, because of knee pain. The inflammation throughout her body had settled into her knees. That's where she was feeling the pain, but in reality it was affecting her whole body.

Over-the-counter pain relievers are hard on your digestive system and liver, often causing indigestion and heartburn, while not really curing the problem. If you have exercise-related pain, you need to space out your intense workouts and focus on low-level exercise while your body heals. You may need several days to recover after intense exercise or competition. You also need to reduce the heat in your body by taking a cold shower. Lingering postworkout heat contributes to bodily inflammation. You can take your usual shower and just finish with cool water for the last couple of minutes.

If you are having joint pain of any kind, you should also think about your diet. This is an important marker for inflammation as well. Think about restricting heavy sauces, spicy foods, and sodium in order to help clear the heat from your body. A clean diet of lean proteins, lots of vegetables, and a few whole grains will reduce toxins and cool your body from the inside out. If you compete, you need an especially clean meal just before and just after competition. This is true of any athlete, not just the ones over the age of sixty. You want clean fuel in your body before and after you push it to extremes. Over time, my patient was able to take fewer pain pills and still enjoy a high level of competition. I consider her an example of the best way to age gracefully while still doing everything you enjoy.

In the Vedic texts, there is an old story about a king who had gained a lot of weight. He was eating too many rich foods and feeling too lethargic to rule. He had no energy and wasn't able to concentrate. He summoned his personal Ayurvedic physician to diagnose his condition and offer a cure. The doctor examined him and told him to set down his crown and leave the palace. The king was told to go live in a distant village. He was to live like the villagers, to eat what they ate, and to do what they do all day. The doctor said to him, "If you live like this for three months, you can come back and be king." Incredibly, the king agreed. He spent three months eating plain food and digging wells all day, and when he returned, he had regained his old vigor and physique. This hard luck sabbatical probably also gave him some much-needed perspective on his old habits.

Now, I'm not going to tell anyone to live in a hut and dig wells all day. But I do often say to people that if they change their schedule to put the body's needs first, their lives will change. Exercise is one of the body's basic needs. It is a pillar of good health because it unites the brain and the heart. It keeps the metabolism humming along. And the truth is that modern life is a lot like that of an ancient king—too many rich foods; too much lying around the palace. In the next chapter, I'll talk about how to find the best exercise for your body type. Armed with this advice, you can get up and get going as nature intended.

9 The Right Exercise for You

When people think about exercise, they usually think in terms of building muscle or losing weight. So many people count the minutes on the treadmill or mark time at the gym as though it's a punishment for bad eating behavior. Exercise is so much more. It's a chance to breathe fully, to work and build muscle, and to burn fat if you are doing it at the right time. You can also do too much exercise and put yourself at risk of inflammation or injury if you are doing exercise routines that are too intense for your body type.

One of my patients is a banker who puts in long hours at the office and has so many meetings during the day that exercise had become impossible for her. At our first meeting, she told me that she hates to exercise, yet she was forcing herself to do long workouts on the weekends to try and stay fit and lose weight. But I had to tell her that this is the great myth about exercise. You can't lose weight just by exercising. In fact, you are probably burning more energy during a good night's sleep than you are at the gym. She was astonished. "So I can just stop?" she asked. Well, no. Not quite.

We all need exercise every day, but we don't all need the same amount or the same level of exertion. What's more, some people do best in group classes, where they can socialize and stay focused. Others need solo walks or runs to clear their heads. A few people need competition or high levels of exertion, but these are the exception. No tendency is better than another. The question is: What kind of exercise do you need? Only when you know your body type and how you respond to exercise physically and mentally, can you choose the right fitness program for you.

To find out how your body responds to exercise, you can take this short quiz. What we are looking for is your natural response to exercise from your own experience. For each of these questions, find the answer that best describes you:

1. **When you were a teenager, how much exercise did you get?**
   a. I did some dance classes, but no formal exercising. I was rushing around doing other things.
   b. I loved team sports. I was always outside playing some kind of game.
   c. I was not that active. I liked socializing, reading, or hanging out.

2. **If you played organized sports (or even kick the can) as a kid, what did you like best about it?**
   a. Every game was different and interesting. I liked chatting during the game.
   b. I loved winning and seeing how hard I could push to win.
   c. I liked the social aspects. It was a great way to hang out with other people, and there were often snacks afterward.

3. **How much time do you set aside for exercise every week?**
   a. It's different from week to week, depending on how motivated I feel.
   b. I like to stick to a regular schedule, at least three times per week.
   c. I'm not that into the exercise scene.

4. **What keeps you from going to a gym regularly?**
   a. I get bored at the gym, and I'm not always in the mood for it.
   b. I've found more intensive exercise in other ways, such as distance biking, running, or competition.
   c. I feel self-conscious there because I don't like the way my body looks.

5. **When you do go to the gym, what is going to get on your nerves the most?**
   a. I have to have music or something to read or someone to talk to or I lose interest.
   b. Someone using the machine that I want to use. I don't want anything slowing me down.
   c. Being surrounded by people who are obsessed with their fitness and their bodies.

6. **When you've tried working out in the past, what got in the way?**
   a. When my workout feels like a rut, I start skipping days.
   b. Work deadlines are the only things that get in the way.
   c. I lose my motivation to get to the gym if there is anything else I'd rather do.

7. **What do you use to track your progress?**
   a. I don't like to keep track of reps or miles or anything, so I count the number of days I go or the number of classes I take in a week.
   b. I need to log my own progress. I use a notebook to track reps, miles, etc. I like to compete with myself.
   c. I don't want to have to track my progress. That sounds obsessive.

8. **If you have taken fitness classes, what was your experience there?**
   a. A class is so much easier than trying to build a workout on your own. I don't mind being told what to do and I know when it's going to end.
   b. I don't like to be told what to do. It seems pointless to work out in a group where everybody has to do the same thing.
   c. I avoid group classes because I don't want to have to compare myself to other people.

9. **If you've tried some intense workout programs, have you had any problems?**
   a. Yes. Intense workouts are great until I get injured. And then I have to stop exercising for months.
   b. No. The more intense the better. If I'm not struggling, it's not working.
   c. Yes. It's too fast-paced for me and I can't keep up.

10. **If you engage in exercise that's more intense than usual, how do you feel afterward?**
    a. Exhausted and wrung out. I need a nap.

b. It resets my emotions; it takes the edge off of my anger and frustration.
c. I feel lighter and more focused.

11. What is it about your daily routine that gets in the way of regular exercise?

a. I'm overscheduled as it is. So the workout drops off my to-do list.
b. Exercise is always a priority. I'd rather exercise than eat lunch.
c. My schedule isn't the problem. Motivation is the problem.

12. If you have to skip working out for a few days, how do you feel?

a. Not that different. Unfortunately, this makes it easier to pass on exercise.
b. Frustrated and fearful that I'm losing my level of fitness.
c. Dull and heavy, especially in the morning. I feel bloated.

13. If you miss your normal workout time, what do you do?

a. I put it off until I have free time, and if I get to the end of the day I just skip it.
b. I want to work out no matter how late it is, so sometimes I'm hitting the gym at night.
c. If I can't get motivated at the usual time, I just skip it. I don't pretend I'll do it later.

14. Do you think of exercise as your primary weight loss tool?

a. Yes, but I don't need to lose that much weight.
b. Yes, I know I can eat more if I've hit the gym.

c. No, but I know I should work out more if I want to lose weight.

15. The best kind of workout is one in which I:

a. Feel energized but not exhausted.

b. Sweat and struggle enough to know that I'm getting fit and toned.

c. Move from feeling reluctant to feeling that I've accomplished something.

16. What I'm most concerned about when I start a new exercise program is:

a. Am I going to struggle with nagging pain or get an injury?

b. Is this going to be intense enough to produce results?

c. Am I going to be motivated enough to continue?

17. How would you describe your body type?

a. Naturally short or thin, with smaller bones at the wrists and ankles.

b. Strong bones and good muscle tone.

c. Big-boned and sturdy.

18. When people compliment you on your admirable traits, they say that you are:

a. Creative and intuitive.

b. Intensely focused with real vision.

c. The calm in every storm and a great listener.

19. When people are critical of your nature, even if it's unfair, they say:

a. I don't finish what I start.

b. I'm unforgiving of mistakes.
c. I can't or won't make a decision.

TOTAL: A: ____ B: ____ C: ____

To score the quiz: Add up the answers. If your answer for most of the questions was A, I would say you are a variable exerciser. If your answers were mostly B, you are a strong exerciser. And if you answered mostly C, I would say you are a light exerciser.

Variable Exercisers

If you answered A to most of the questions, you are a classic variable exerciser: someone who wants to be fit, but often finds that something gets in the way. In Ayurvedic terms, your body type would be called "vata," which means air. And like the air, you are a shifting, swirling mass of ambitions, plans, and contradictory impulses. In truth, you burn most of your energy through fidgeting, talking, and rushing from one task to the next. You've got a lot of plates spinning, and you rely on your natural creativity and intuition to keep them all in the air. As a result, you probably exercise in streaks. You have found exercise routines or fads that worked for you for a few weeks or months, and then when you get bored or suffer an injury, you forget to exercise.

When you were younger, you may have found that dance was a great physical outlet because you could go to classes and engage in something active that was also creative. In your twenties and thirties, you may have found that training for a road race could capture your attention because there was a set goal in mind. If I could peek in on you at the gym, I would see someone who is listening to music and reading a book while simultaneously watching the bank of television screens above. You would be doing anything to kill the boredom of being on a treadmill or counting reps at a weight machine. Boredom is your primary challenge when it comes to exercise, and you need exercise that can engage your mind along with your body.

Exercise is also a struggle sometimes because of your variable energy levels. Your energy is like the wind. It comes in great gusts before it falls still without warning. Variable exercisers who engage in intense workouts may find themselves wrung out and emotionally exhausted afterward. While other body types can be energized by an intense workout, you tend to feel spent, and it may take you ninety minutes or more to rebuild the energy you need to focus on your other goals. That's another reason why variable exercisers sometimes feel that exercise is a drain on their schedules. When you do too much, you pay the price afterward. You can also put yourself at risk for injury. If you've had nagging injuries in the past, you were likely doing too much. Your dry joints can get inflamed and sore, especially if you don't have enough healthy oils in your diet.

The good news is that with this body type, you don't need intense exercise to get or stay fit. As long as you are doing enough to breathe deeply, get your heart rate up, and move your whole body, you are doing fine. People of this body type sometimes struggle when they hire personal trainers who are too intense. A lot of fitness programs are designed by people with a very specific body type and a great affinity for struggle and sweat. These things are a tonic for some body types, but not for you.

Here are the pointers I offer to people of this body type:

Stay mentally engaged in your exercise. You need to find exercise that engages your mind as well as your body. That probably means a fitness class where you can move to music or follow instructions. Yoga is also a great choice because it can calm your racing thoughts and deepen the mind-body connection. By contrast, reading and watching TV while on the treadmill is a bad idea because you are not paying attention to the way your body feels during exercise. This puts you at greater risk for injury and exhaustion. Instead, you need to find a variable program that's not too intense. You can walk or do isometric exercises on some days, do a yoga class or dance class on other days, and rest on the remaining days. Variation is the key to motivation.

Do exercise first. I tell everyone to exercise before breakfast, but everyone needs to do this for different reasons. As a vata, you need structure in your day. You need a set schedule so that you can achieve your goals. Variable exercisers tend to put off this important task, but by putting it first, you are making it a priority. Find something that you can do every day, and then do it every day. Also, you may think that you wake up easily in the morning, but in reality, your mind is a jumble of conflicting goals. Morning exercise is a kind of meditation. That deep breathing and gentle sweating will put your mind and your body on the same page. You will be able to focus more easily on work with a little exercise under your belt.

Less is more. Look for grounding exercises that increase flexibility at a more relaxed pace. This means yoga, Pilates, tai chi, walking or hiking, or cycling. Don't worry about whether you are sweating enough or straining enough. If you do more rigorous exercise, it's important to check in with yourself during and after every workout. How do you really feel? Ideally, you should feel energized and calm during your exercise routine and afterward. If instead you feel light-headed, depleted, or suffer from muscle cramps, you have probably done too much. Interval training is great for this body type because you have short bursts of high intensity followed by much longer rest periods. You don't want to spend large portions of your workout gasping or struggling. If you do, you are sending your body into high stress mode, and that's the opposite of fitness.

Make time to warm up. Other body types can jump into an intense routine with minimal warm-up, but not you. You need to stretch and spend time doing deep breathing while you ease into a workout. This is particularly true in winter, when variable exercisers struggle with chills and poor circulation that keeps their hands and feet cold. Make sure you use a lot of stretching and at least five to ten minutes of walking or jogging lightly to get your circulation moving and your lungs working before you dive in.

Hydrate all day. This body type is especially prone to dehydration, in part because you are rushing around all day long. You also probably forget to eat and drink when you get engrossed in an activity. During and after workouts, this dehydration can get critical, so you want to be drinking water before, during, and after exercise. You will also probably need to increase the natural oils in your diet by eating more nuts and seeds. This is especially true as you move past the age of forty. This won't put you at risk of weight gain. By contrast, it will lubricate your joints and make your skin radiant. Many times, I've worked with variable exercisers who can't understand why they have no energy, but when I look at their intensive workouts and fat-free diets, I see the problem right away. If you are struggling with fatigue after increasing your workout routine, dehydration is the likely cause.

Body Type Injuries

Knowing your exercise body type can help you avoid injury. Your body will react to exertion differently than others.

Vata: You have a naturally thin frame with small bones, slender muscles, and ligaments that tend to be dry. Exercise will help define your muscles and strengthen your bones. Too much intense exercise will cause your bones to creak. You know you've done too much when your knees, ankles, or shoulders make a crunching sound when you move them. Advanced hot yoga or CrossFit can leave you wrung out and put your body in crisis. You don't want to be in any kind of class or program that's competitive or that pushes you too hard because you can easily strain muscles and pull ligaments. All of those DVDs for extreme athletes are a disaster for your body. Avoid exercise instructors who urge you to push through the pain or feel the burn because they don't understand your body type or your fitness goals. This is especially true as you get past age forty. Gentle, mindful exercise is the goal.

Pitta: You have strong bones and muscles and a strong competitive drive. You need intense exercise because it gives you emotional equilibrium. Your problem is the heat in your body can easily inflame your tissues and joints. You often push through the pain in order to finish a workout or a game and this can lead to injury. When you aren't connected to your body during exercise, you won't know that you're hurt until it's over. After the workout, you'll find nagging pain in areas that don't seem connected to exercise, including your back. You know you are doing too much when you develop a dependence on over-the-counter anti-inflammatory medications. If so, you are failing to cool your body adequately between workouts. After intense exercise, take your shower as usual, but finish it with cold water for the last couple of minutes. This will reduce inflammation throughout the body and speed your recovery before the next workout. This is particularly true as you age, when overuse injuries and nagging injuries can interfere with your love of exercise. As you age, you want to add moderation to your exercise routine. Take more days off, and include more days with modest exercise instead of going full bore.

Kapha: You may be a latecomer to intense exercise, but it will do you more good than anyone else, and you have the stamina to do almost anything. You have strong bones and muscles that can handle hard work. Sweaty workouts purge your body of toxins and clear your mind. The trouble starts when you cut corners in your workouts. Using improper technique in weight lifting, hot yoga, or Zumba will spell your doom. If you are carrying a lot of extra weight, you are already straining your joints. When you move quickly, you are releasing a lot of kinetic force that can pull or tear muscles. Kaphas are more likely to face injury from improper technique than anything else. Be careful with any new exercise routine and build up slowly in weights, reps, and minutes on the treadmill. Any time you feel a sharp pain or twinge during your workout, stop and check your stance. If you are lifting weights, you may need to reduce the weight until you have the movement perfected. You don't want a sudden injury to derail all of your progress.

Strong Exercisers

If you answered B to most of the questions, you are a strong exerciser, which means you probably live by the motto "no pain, no gain." And that's okay, because for the most part, it works for you. You have probably spent most of your life putting exercise near the top of your list of priorities. No other activity shapes your body and clears your head like a daily trip to the gym, where you can do your weights and cardio. Or perhaps you have become addicted to CrossFit or one of the many DVD programs that promise to shred your muscles or make you ready for competition. In Ayurvedic terms, you would be a pitta, and "pitta" means fire. In your exercise routine, you crave intensity, heat, and progress. When you are not at the gym, you are on a bike, on a tennis court, or swimming laps. The best thing about working out is knowing that you are getting stronger and more capable every day. If I saw you in the gym, I would recognize you by your intensity. Either you are moving from weight bench to weight bench, or you are logging progress in a notebook to make sure you are advancing in strength and ability in every workout. Scheduling workouts is no problem for you, just as keeping a regular routine in your daily life is a priority. Even on vacation, you are looking to get as much out of every day as possible.

You've always been active. As a kid, you were the one running from ball game to ball game, and you may have felt that the best way to relax was to shoot some baskets or play a pickup game. In your teens, you joined a lot of teams and enjoyed pushing yourself to excel physically. In your twenties and thirties, you may have cycled through various fitness crazes, always looking for something that would help you build strength and stamina. Strong exercisers usually have a large appetite because their digestive fire is as intense as their personalities, and so you may have felt that exercise was a great way to balance your love of food.

For many strong exercisers, activity is a weight-loss regimen. Strength training builds muscle and burns the fat that would otherwise accumulate from an erratic diet. Having a strong body and an equally strong metabolism works well as a weight-loss solution until middle age. At that point, many strong exercisers find that their metabolism shifts and their bodies no longer build muscle at the same rapid rate. If they are unlucky enough to suffer an overuse injury from intense training, they can find that they put on weight rapidly and have no real way to lose it. I urge people with this body type to get their diets under control before they turn forty, so that they can stay trim and age more gracefully.

If you have this body type, here's what you should keep in mind as you exercise:

Leverage your love of competition and intensity. Because you have a larger frame and more muscle mass, you need higher-intensity workouts to keep your muscles toned and to discharge all of your pent-up energy. You also have a stronger natural drive and love to feel challenged, so a difficult workout helps reset your emotions. It takes the edge off of your frustration and your inability to control everything in life. Try to engage in sports that combine competition with exercise, such as basketball, racquetball, or martial arts. Any activity that requires you to track your progress helps you stay motivated, including CrossFit training or distance biking. You can excel at winter sports because your body isn't affected by the cold. Stay away from golf if it increases your level of stress. Be careful, too, about pursuing adventure sports, such as BMX biking, extreme skiing, and snowboarding, which can also take a toll on your body.

Avoid lunchtime workouts. You are often overscheduled, which means that you are tempted to exercise later in the evening or at the noon hour, instead of eating. Noon workouts push your blood flow away from your digestive tract, which will upset your stomach when you do eat. You have more acid in your stomach at mealtimes than other body types. If your mealtimes come and go and you don't give your body real food, that acid will sit in your stomach, where it will be reabsorbed and create more heat and inflammation in your body. Morning workouts will offer you an emotional reset first thing, so that you can face the day calmly and with greater focus.

Log your progress. If you are getting back into exercise after a pause, the best thing you can do is use a spreadsheet, worksheet, or small notebook to keep track of your minutes on the treadmill, your reps at the gym, or the miles you have run. You can also track your food and water intake if you like. You are detail-oriented, and you love routine more than any other body type. Giving yourself written proof of your progress will keep you focused and energized during your workouts.

Stay cool. You have a natural fire, meaning that your body temperature runs high. You sweat a lot when you work out, and that's fine, but you need to make sure you don't get overheated. High heat increases inflammation and puts you at risk for injury, so be sure to hydrate, and you may also benefit from a cool shower after an intense workout. This natural fire also speaks to your emotional intensity. You can and do push yourself too hard in competition, and that's when you get hurt. You sometimes try to lift too much weight, or force your body to do things it can't. The antidote to this is staying cool mentally as well as physically: keep your mind sharply tuned to your body while you work out. Stay curious about how you feel during each exercise. Swimming is an ideal exercise for this body type, especially if you are dealing with nagging injuries, because it keeps the body cool and takes the pressure off your joints.

Check your recovery period. After intense exercise, pay attention to how your body feels. In the first few minutes, you may want to relax and check in with your body. This cooling-down period becomes even more critical as you age. After age fifty, you will need to space out your workouts so that you don't overdo it. If you are taking pain medication every day, that means your workouts are too frequent and intense and you aren't cooling down enough at the end.

Try relaxing exercises. When I tell people with this body type to go to a yoga class, they often rebel. They don't think that poses or slow movements can be beneficial. They think yoga classes are a waste of time. I understand your love of sweating and straining, and yet everybody needs balance. You need intense exercise, and you also need relaxing exercise that builds the mind-body connection. That might mean a walk on the beach or in the woods a couple of times a week. It might mean trying a yoga or tai chi class.

Light Exercisers

If you answered C to most of the questions, you may struggle to get enough healthy exercise. You may go to the gym intermittently, but it's never been a priority. In Ayurvedic terms, your body type is called a kapha, or water type. You are too easygoing to get obsessed with your body or to get caught up in competitive activities. And that's a good thing. But you do need exercise. Without it, your body takes on excess fluids and you feel sluggish and dull. Of all the body types, yours responds best to exercise because you rarely face injury and you never tire. An intense workout will fill you with vitality and give you a healthy glow.

This may be a revelation, because a lot of light exercisers think of themselves as overweight. Your body type is one that built bones and tissues early in life. And as you sailed through the teen years and young adult life, you may have struggled to lose weight. And you may have a tendency to put on weight, regardless of what you eat. In childhood, you were more interested in socializing or reading than in athletics. If you became self-conscious about your body as a young adult, that probably kept you from joining a gym or taking fitness classes. In a culture obsessed with thin bodies, it's hard to feel comfortable if you are different. You may also feel intimidated by starting an exercise routine if you feel out of shape. You probably love the social aspect of the gym and still struggle to stay on track with a workout. You might try walking outside or on a treadmill so that you can feel the power of exercise on your body and your mental clarity without worrying about how you fit in with a gym culture.

Here are some guidelines for choosing the best exercise plan for you:

Do your workout early. Remember that activity energizes you, particularly if you do it in the morning. If you have trouble getting going and staying focused in those first hours after you wake up, exercise can change all of that. It can make you feel lighter and happier. You need to tap into your energy stores to get your mind going. In particular, you need to work your lungs. Do anything that causes you to breathe deeply. Some of my patients say that exercise is better than coffee in the morning, because it makes them feel so focused and alive.

Go the distance. This body type has the most endurance because you have larger bones and more tissues. You can work your body endlessly without worrying too much about injury or exhaustion. That means you will do really well with power walks, hiking, rowing, distance running or biking, or anything that requires steady, sustained energy. Even if you start with walking, you can build up longer distances, then alternate with running and walking. Eventually, you can add weight training or a higher intensity class, such as spinning or aerobics. Once you get past your inertia, you work harder and longer than anyone in any fitness class.

Be sure to sweat. This body type tends to retain water, which is one reason why you may feel bloated, congested, or sluggish in the morning. You need to sweat out these fluids with exercise. In some cases, I've advised people to do a twenty-minute turn on the treadmill first thing, followed by a sauna. You can get a sweat going with exercise and then keep it going in the sauna, which will help clear your body of excess fluids. It's okay to start with a half hour of brisk walking if you haven't exercised in many years, but as you continue to exercise, keep ramping up the intensity so that you can break a sweat every time.

Observe good technique. If you are new to exercise, you can get the full benefit by doing simple walking every day. Even this amount of effort will cause you to breathe deeply and break a sweat. Continue to push yourself to do more week by week, but always observe good technique. If you are carrying extra weight, you don't want to hurry into unusual movements or try to lift too much weight because this will strain your joints. Hot yoga is good for you, but never rush the postures. CrossFit is good for you, too, as long as you don't rush until you've learned each exercise. It's best to push yourself mindfully, so that you don't get hurt.

Connect to your body. It's important to notice how you feel during your workout and afterward. Many people with this body type have forgotten how great they feel while engaging in intense exercise. They have forgotten how light they feel and how clear their minds get. Make time to notice the natural high that comes from exercise and you will be more likely to keep doing it.

While you must eat and sleep in order to stay alive, exercise is a different kind of habit. It's something you must choose to do every day to maintain your body's vitality. Now that you understand your body's unique exercise needs, you can cultivate a habit that will enhance your diet and the quality of your sleep. You can find something that nourishes your energy level and something that is also fun to do. This is particularly true in a time when so many people work at desks all day and relax on the couch at night.

Now that you've learned how to sleep more deeply, eat more mindfully, and energize your body through exercise, we'll talk about how your body's circadian rhythm changes through the seasons.

10
Your Body Through the Seasons

Just as the body relates to the changing light during the course of a day, the body also changes in relation to the seasons. Your body has to react to the changing amount of light and the changing temperature during a calendar year. Whether you realize it or not, your blood pressure and blood cholesterol levels change seasonally, and so does your body's need for food, exercise, and sleep. I often ask patients how they alter their diet, sleep habits, or exercise depending on the time of year. Usually, they say that they don't change a thing, despite reporting that they have a starkly different emotional and physical reaction to the changing seasons. They say they love the summer, and when I ask why, they say it's because they get to go on vacation. When I ask how they feel during the hot summer months when they aren't on vacation, they realize that hot weather makes them frustrated. Or they say that the winter is great because of the holidays, but in reality they suffer from congestion and sadness during most of the winter months. These things matter because you can change your schedule to help you cope with the physical challenges that each season presents for your body and your emotional health.

Even though you notice that it's getting colder in winter or that the days grow longer in spring, you may not be altering your schedule to reflect these changes. When people tell me they have a fruit smoothie or berries and yogurt every morning all year, I have to ask: Why are you doing this? They say it's natural. Fruit is good, right? But blueberries are a cooling fruit. Mother nature gives these to us in summer to help counteract the heat, and at a time of year when the body can better process fructose. Others tell me that they give up on exercise in the winter, because they can't run outside or because it's tough to go to the gym in the morning when it's still dark. But the winter months are exactly the time when your body needs the warming and energizing effects of a daily workout. This is exactly the time to give up on cold sandwiches or cold salads for lunch because they won't warm and nourish your body. Your diet, exercise, and sleep need to adapt to the changing seasons. And you have to pay attention to how your body feels in each season in order to make the right changes.

Think about it: in the heaviest days of winter, when the snow piles up around you and the heat is going all day, you may feel scattered and dehydrated. The short days may interfere with your mood and your enthusiasm for working out. You may reach for sweets or other comfort foods to get you through, and you likely gain weight. In spring, nature takes all of that melting snow and rain and we see tremendous natural growth, yet there may still be a kind of heavy or lazy feeling inside the body. From late spring to early summer, you see the greenery all around, the animals and birds are out and making noise, and your energy level increases along with your mood. In the summer, you want to be outside and active as much as possible, unless, of course, it's too hot. You might feel that you can eat more without gaining weight, but in reality, you are probably eating more fresh produce and getting more fiber than you do in other seasons. And yet on the hottest days, you may feel languid or frustrated. In the fall, after the harvest, you begin to cycle back into that dry season as the days get shorter. You might feel a quickening of ideas and vision as the days become shorter and cooler, or you might become sad and lethargic as the natural world begins its cycle of hibernation. All of these should be clues to help you change your diet and schedule. You might think that these things are all a function of the outside temperature, but there is much more at work here.

The Body Through the Seasons

For years, researchers have observed that the body changes throughout the seasons. Blood pressure goes up slightly in winter and down in the summer.[1] So does your cholesterol level and level of C-reactive protein, a marker for inflammation.[2] Given all of these factors, it's probably no surprise that cardiac incidents, including heart attacks and strokes, peak in the winter months; so does the incidence of type 1 diabetes.[3] In fact, death from all causes peaks in winter, but then so does conception, which remains high throughout the spring. We tend to take in more calories in the fall and winter, and people report that they feel hungry even after they overeat.[4] We naturally reverse this pattern in the spring, feeling less hungry and eating less as a result.

Even the brain changes throughout a calendar year. Serotonin levels peak during the summer months,[5] while dopamine levels go up in the fall and down in the spring.[6] Falling serotonin levels in the fall and winter may be a contributing cause to seasonal affective disorder.[7] But the brain changes in other ways as well. Using functional MRIs, researchers have noticed that different parts of the brain are more or less active at different times of the year, which suggests that the ability to learn, think, and remember may also have a seasonal component, but these things are harder to measure.[8]

The immune system also is less active in winter, particularly the genes that suppress inflammation. The research on the seasonality of our genes is new and ongoing. One recent study shows that about one-fourth of our genes change through the seasons; some are more active in summer, while others are more active in winter.[9] These changes affect our immune systems, our fat cells, and the composition of our blood. These changes seem to be triggered by the changes in the light cycle. And this implicates the hypothalamus and the tiny bundle of nerves in the suprachiasmatic nucleus that registers light and sets the body's circadian rhythm. It seems that our bodies' functions are tied to the seasons and not simply to the daily diurnal rhythm of light and darkness.

The Seasons of Ayurveda

All of this research supports the Ayurvedic view of how the seasons affect the body. In Ayurveda, the three dominant seasons—spring's growth and moisture, summer's hot productivity, and late fall/winter's dormant dryness—correspond to three doshic energies: kapha, pitta, and vata. The changing seasons of nature influence your body in so many ways. Everyone has a season in which they feel great and probably one in which they struggle. By understanding the energy of each season, you can prepare your body to stay balanced and healthy throughout the year.

Kapha Season (Late Winter to Early Summer)

Kapha is the season of growth and moisture. It becomes active at the winter solstice, becomes predominant in late winter, and wanes at the end of spring. This is the season when the body is primed for growth, which can be good and bad. It's a time to start thinking about the exercise you need to build your muscles and strength. It's also a time when the body naturally wants to shed toxins and fat, which means it might be generating more mucus and you may be more susceptible to colds and seasonal allergies. To embrace this season, think in terms of what your body needs:

Sleep: As daylight comes earlier, you should be waking up earlier. Early in this season, you will still be getting up in the dark for work and coming home in the dark, but you want to get as much natural light as possible, which may mean getting outside for a short walk before work and another one after lunch. This natural light will help you lighten your mood and open your body to the spring energy. You may find the start of this season corresponds to heavy sleep and a dull lethargy in the morning. As the season progresses, you may find that your sleep gets lighter. That's natural. As long as you have a steady diet of natural light and some exercise, you can keep insomnia at bay.

Exercise: At the New Year, you should be thinking about heading back to the gym—and not just because of New Year resolutions. This is the season of lethargy, and exercise is the best medicine to combat this feeling. Concentrate on a morning routine that builds strength and flexibility. Don't overdo it! You can do some stretching and weight training, along with cardiovascular exercise. And increase your level of exercise throughout the season. This is the prime time to use exercise as a weight-loss and body-sculpting tool. The kapha season is a time of building tissues. Your muscles are primed to gain strength and endurance now. You might want to follow each workout with a sauna that helps the body shed moisture. You won't need saunas as much after the spring equinox, when the lengthening days signal a waning time for the kapha dosha.

Diet: Your metabolism has been hibernating through the winter. You have probably been eating more meats, high-fat foods, and root vegetables (and holiday sweets, if you are being honest) while bundling up against the cold. Although the cold, dry weather will continue in many areas through January and February, along with indoor dry air, your body is getting ready to accumulate water and shed toxins. It's no longer in the mood to store fat, so it's time to introduce the spring diet even before the first springlike weather shows up. Beginning early in the New Year, you are looking to reduce heavy, oily foods and introduce spring vegetables, such as asparagus and sprouts, as you prepare for warmer weather. You will feel less hungry than you did in the fall and early winter, so this is a good time to cut back on your portions and reduce fats. As you get into February, you want your diet to look like this:

1. **Bitter greens.** Transition from root vegetables, such as carrots, potatoes, and beets, and start eating more green

vegetables. This is the season for spinach, cabbage, broccoli, sprouts, asparagus, onions, and peas. You can also eat onions and onion sprouts.

2. **Leaner meats.** You want to reduce heavy, oily meats you may have indulged in over the winter. It's okay to reduce your protein at this time and get more of it from lean meats, eggs, nuts, and tofu. Reduce oily, deepwater fish as well because you don't need as many oils in your diet at this time.

3. **Less dairy.** Because your body may be retaining water and producing mucus, you want to stay away from dairy products as the season progresses. Dairy products contribute to seasonal colds and allergy symptoms. The one exception is warm milk, which you might be using to help you feel drowsy in the evening, or warmed for your coffee in the morning. Warm milk contributes less to congestion, particularly when seasoned with cinnamon or other spices.

4. **Dried fruits.** There aren't a lot of natural fruits in season during the spring, and so you want to reduce these as well, which means it's not yet time for berries in the morning or fruit-based smoothies. The earliest berries come into season at the very end of the kapha season. If you need fruit, you can have dried fruits in small quantities.

5. **Lighter oils.** This is a good time to lighten up on cooking oils as well.

6. **Warming spices.** You want to reduce your salt intake while the body is accumulating moisture, but all other spices are really good for you. Hot spices, including peppers, open up the nasal passages and cause flushing that will help purge toxins.

Kapha Balancing Tea

This tea helps the body prepare for the spring season. It includes warming spices that promote cleansing, aid digestion, and give you energy. They also help to regulate blood sugar and to naturally flush extra fluids from your system. You can take it at the start of the New Year until early June.

4 cups water
1 tablespoon cinnamon or crushed cinnamon sticks
¼ teaspoon turmeric
1 tablespoon natural sweetener (honey, rice syrup, agave)
1 teaspoon grated ginger

In a medium saucepan, boil the water, then add the cinnamon and turmeric. Add the sweetener and ginger and boil for 2 minutes. Strain and serve.

Pitta Season (Early Summer to Mid-Fall)

Pitta is the season of heat and productivity. Its influence emerges at the spring equinox, but it becomes dominant in early summer and wanes at harvest time. After the spring equinox, the days get longer and warmer. Rain and humidity become intermittent and mother nature is alive and vibrant. The plants are doing their thing: growing and producing fruits and vegetables. This is the time to follow nature's lead and be more active, work on your goals, and get outside. The body is best suited to take in light and activity and can better tolerate a more varied diet. Here's how to take full advantage of this season:

Sleep: The brain is most active during this season, so you can sleep less while still feeling refreshed. That does not mean staying up late, but getting out of bed at first light will be easier, allowing you to take advantage of the longer days. That may mean waking up at five or five thirty during the longest days, but you will want to get up and soak in this extra light. Even if full darkness comes on after eight thirty, remember to put your electronics away as soon as it gets dark. Your body needs a robust reminder to wind down and get ready to sleep when the days are longest.

Exercise: This is the season to be outside. No matter your body type, endurance peaks in this season, so you can be more active outdoors for much of the day. Going on long walks or hikes is ideal. Be careful about overheating during workouts, and make sure you get lots of fluids on hot days. Because of the heat and humidity, inflammation is a chief concern in summer. If you find yourself with nagging pain, dial back on your exercise and opt for low-impact movements and stretching for a few days.

Diet: This is the high season for eating fresh fruits and vegetables. Your body is getting lots of light and activity and, at the same time, you aren't feeling as ravenous as you do in the fall and winter. This is the time for salads, fruits, and more carbohydrates. Favor foods that are bitter, astringent, and sweet. By late June, your diet should focus on:

1. **Raw vegetables.** Think of eating a wide variety of seasonal vegetables at this time, either raw or cooked. If they sell it at the local food stand or farmers' market, then it should be on your plate. You can eat beans and a few potatoes during this time, because your digestion is strong enough to handle complex natural starches. Even so, you

want to buy foods that are locally grown and ripe because they contain the energy of the season. Stay away from other root vegetables. Also, dial back on the hot peppers, particularly on hot days, because they cause the body to flush and sweat.

2. **Less meat.** You can reduce your protein intake overall. Eat lean meats only, which is okay because you probably aren't as hungry for heavy meats and high-fat meats when the body wants to be light and active. (If you've ever wondered why you feel heavy and dull after eating traditional cookout fare of burgers and hot dogs, that's why.) It's a great time to grill lighter fish, lean proteins. Even so, eat these in moderation. Your body needs colorful natural foods to keep your energy high.

3. **Moderate dairy.** You can increase your intake of dairy during these months if you don't suffer from seasonal allergies or summer colds. If you do have congestion, ditch the dairy products.

4. **Vibrant fruits.** This is the high season for fruits, so you can eat them at every meal if you want. Your body is better able to handle their natural sugars, and they also offer a cooling benefit to the body.

5. **Healthy oils.** Though you likely won't crave oils and heavy foods, you don't need to worry about reducing your oil intake at this time. Use the same amount of oil that you did in spring cooking. Your body still needs lubrication, regardless of the season, and healthy oils help kindle your digestive fire.

6. **Savory spices.** You can use all savory spices at this time, but you want to reduce mustard, cayenne, and chili peppers

because they cause your skin to flush and they create heat in the body. At this time of year, you want to keep the body cool.

Pitta Balancing Tea

Prepare this tea between June and the end of October to help the body deal with the heat and activity of the summer season. These herbs help cool the body down. They can kindle the digestive fire without heating the body. These same herbs reduce inflammation, which is especially problematic during this season.

- 4 cups water
- 1 tablespoon crushed or chopped mint leaves
- 1 tablespoon crushed or chopped spearmint leaves
- ⅛ teaspoon saffron fibers
- 1 tablespoon natural sweetener (honey, rice syrup, Sucanat), optional

In a medium saucepan, boil the water, then add the mint and spearmint leaves. Let it steep for 3 minutes. Remove the saucepan from the heat and let it cool before adding the saffron and sweetener, if you choose. Strain and serve cool, but not with ice.

Vata Season (Late Fall Through Winter)

At the autumn equinox, nature takes a rest. The leaves have fallen and the vegetation has gone dormant. Animals prepare for hibernation and the weather becomes cold and dry as the days grow short. Vata is the dry season of hibernation and reflection. It becomes active at the fall equinox and dominant during late fall, then wanes at midwinter. You may notice that you are sleeping more heavily and that you feel more scattered and less social.

This may be the season where you wake up for work in darkness and then find that it's dark when you come home at night. The body becomes more naturally dehydrated, even if you don't need to run the heat all day long or bundle up to stay warm. Becoming dehydrated will make you more lethargic and affect your mood and your thinking. This is a season to keep drinking water and herbal teas all day and to think about increasing the amount of oil in your diet. Remember that your skin is the largest organ in your body, and investing in natural oils to apply to your hands and feet is a good way to help the body maintain its moisture.

Sleep: This is the time to make sure you turn the electronics off early enough to rest before it's time for bed. There is a tendency to invest in passive entertainment in the coldest and darkest months of the year. This is the trap of these short, cold days. When it's dark outside, you are focusing on finding ways to entertain yourself. Instead, choose card games or board games, which are slightly more active. You will naturally sleep harder and longer during this season, so it is especially important to get to bed on time and leave enough time for sleep before the alarm goes off. This is the season when darkness stays later in the morning and comes earlier in the evening. Make time at least twice a day to get outside and walk, and work next to a window if you can. You need as much natural light as you can get in order to feel sleepy at night and keep seasonal sadness at bay.

Exercise: As the days grow short, you may not feel as motivated to do long stretches of running or serious weight training.

This is the season to prioritize gentle movements and stretching. Try yoga, Pilates, or anything that allows you to keep your body flexible. You can still run and do weight training, if you are a strong exerciser—a pitta type—but this isn't the best time of year to build your strength or endurance.

Diet: This time of year is most problematic for people trying to lose weight. You may be hungrier than you were in summer, and you may be tempted to overeat. The holiday season is no help here. Use the ginger drink (page 127) to help you avoid cravings and overeating, and be diligent about avoiding flour-based foods and sweets. This is the time of year emotional eating is most tempting and most damaging to your mood. If you notice that you are eating more sweets or drinking more alcohol, you may be doing this for emotional reasons, because of a seasonal depression. This is true even if you are going to parties. Be sure to eat healthily before you go to social occasions and minimize alcohol consumption, which only worsens the winter blues. When the days are shortest, you need to eat your final meal of the day before dark. You also want to avoid foods that dry out your system, including baked goods and dry snack foods, such as popcorn and crackers. Avoid cold foods, including yogurt, cold cereals for breakfast, or raw vegetables. Think in terms of warming foods, particularly at breakfast. By the end of October, your diet should include:

1. **Root vegetables.** There are a lot of good seasonal vegetables to choose from, including winter squashes and root vegetables. You can have bitter greens, if they are cooked as well. Think of soups and stews as your mainstay during this time. They are warm and filling.

2. **All meats.** This is the time to eat meat, if you enjoy it. Meats contain macronutrients that your body can use throughout the year. If you are taking in more energy from beef and fish, you will naturally reduce your intake of refined sugars and flours. Eating more meat doesn't mean eating larger meals. You still want to have small portions of meat surrounded by plenty of vegetables, but more protein at this time of year will sate your increased hunger and prevent snacking.
3. **No dairy.** Reduce or eliminate dairy during the dry months because the combination of dry nasal passages and mucus can make seasonal colds much worse. If you have a variable metabolism, you can have dairy in small amounts, but people with a slow metabolism should avoid it altogether.
4. **Soaked dried fruits.** In the dry months, there are very few seasonal fruits to choose from. You can have apples, especially if they are cooked. But you want to avoid buying berries out of season. Frozen fruits in smoothies are a bad idea during the winter, when your body is less able to handle cold foods and the sugars that berries contain. If you crave fruits, you can have some dried fruits on your oatmeal in the morning, but be sure to soak them first.
5. **Plenty of healthy oils.** This is the high season for using oils in cooking. The body craves natural oils to keep the skin and joints lubricated. As you take in more energy from natural oils, you will want to reduce sweeteners and refined grains.
6. **Warming spices.** You can have more salt at this time of year because it helps you to store fluids in the body. You can also increase your use of peppers and other spices because they help keep you warm.

Vata Balancing Tea

Serve this tea between the end of October and early February. This warming tea will help you stay grounded during the cold, dry season, while keeping you hydrated. These herbs help regulate the appetite, which can become variable at this time and also promote healthy bowel function.

4 cups water
1 tablespoon fennel seeds
1 tablespoon coriander seeds
2 cardamom pods
1 tablespoon natural sweetener (optional)

In a medium saucepan, boil the water. Crush the seeds and pods slightly and add to the boiling water. Boil for 3 minutes before adding the sweetener, then boil for 2 more minutes. Strain and serve hot.

Detoxing Between Seasons

Because your body has to adjust to the seasonal changes in light, it's customary in Ayurveda to do seasonal cleanses to help the body release the energy of the old season and to welcome the new season. In Ayurveda, this is called *ritucharya.* In Sanskrit, *ritu* means "season" and *charya* means "routine or schedule." A formal, multi-day cleanse, called a *Panchakarma*, which involves oil massage, steam, bowel cleansing enemas, and a diet specific to your health needs, can be done at a clinic. But if you don't have the time to do all of that, you can try a more simple digestive reset at home.

You want to do this at or near the dates when the days are shortest, again when the days are longest, and at each of the equinoxes. Each year, the dates would be approximately the twenty-first of December, March, June, and September. You don't have to choose these dates exactly; any time about a week before or after will do. Pick a day that is on a weekend, or one in which you won't need to go to work or deal with a lot of social obligations. This is a time to stay home and take care of yourself. You will need to have ready access to a bathroom for the first few hours.

Start in the morning by eating nothing when you wake. Black coffee or tea only, if you need caffeine.

First thing, take one to two teaspoons of castor oil, which is a simple laxative. You can take it plain or mix it with a little grapefruit juice if the taste bothers you. Refrain from eating anything else until noon. After about thirty minutes or an hour, the oil will begin to do its work clearing out your colon. You are releasing all of the accumulated toxins of the previous season, and all of the food that has been left undigested in your system.

For the rest of the day, you should stick to a liquid diet of clear soups, broth, and herbal teas. If you feel you must eat something in the early evening, make it light, such as a piece of fruit or steamed vegetables. You want to give your digestive system a rest so that it can begin working with renewed energy the following day.

If you take the time to cleanse your body as the seasons change, you will be less likely to suffer from seasonal colds, and your ability to digest new foods will improve dramatically. You should notice an increase in energy in the weeks following this cleanse and a greater ability to make good food choices.

How Does Your Body Feel?

When you know your body type, you need to pay attention to how it behaves when the season matches your own dosha. Though it would seem natural to assume that you'd be at your best when your own dosha coincides with the season, the opposite is actually true. When the season matches your dosha, you are most at risk for being out of balance. If you are a kapha, for example, then the spring kapha season is when you need to be vigilant about a seasonal diet and about your schedule so that you can counteract the effects of this season on your body. But what if you live in a region where the season that corresponds to your dosha is unusually long? This is when you can have ongoing health problems that are difficult to treat.

I met with a lady recently who was living in Idaho, a place with long winters. The spring season can be cold and damp for many months. She came to see me because of chronic sinusitis. She was congested most of the year, and she had a cabinet full of nasal decongestants and sprays. Two doctors had recommended surgery. She had been suffering with this condition for ten years, exactly as long as she had been living in Idaho. Every year she goes to visit her sister in Arizona. I asked how she liked it there, and she said she'd realized that she didn't have congestion in Arizona. Those fifteen-day visits were symptom-free. So I asked the next natural question, which was how she could move to Arizona. She laughed and told me that this would be a very expensive prescription to fill. Instead, she told her husband that she wanted to stay with her sister for two months and see how she felt in a different climate that seemed more compatible with her body. Because she runs a small internet boutique, she was still able to work and earn money in a different location. But at the end of those two months, she'd had a complete remission of her congestion and sinusitis and she had lost eighteen pounds.

How is this possible? She was a typical kapha type, a strong sleeper with a slow metabolism, and that long winter was exacerbating her sinus trouble and her inability to sleep, and she was eating more as a way of distracting herself from these symptoms while not having the energy to exercise. In a new, dry climate, all of these symptoms went away. She breathed easier, she slept better, and she ate better. She walked daily and even started more rigorous exercise, and she was not excited about moving back to Idaho at the end of the two months. Ultimately, her husband agreed to look for work in Arizona and they were able to make the move permanently.

Of course, it's not possible for everyone to consider moving, but by being careful about your diet, exercise, and sleeping routines, you can do a lot to minimize seasonal distress. This is the power of listening to your body and finding out how you feel in each of the seasons of the year.

The first step is to ask yourself how you feel at different times of the year. How is your health in different seasons? How does your body interact with the changes of light and darkness and temperature? In which season are you at your best in terms of health, energy, and mental clarity? In which season do you feel your worst? It's one thing to say that you hate winter because of the cold and the short days, but some people enjoy winter. People whose body temperatures run higher can withstand the cold. They may enjoy outdoor sports in winter. They may have jobs that allow them to get enough natural light. Others struggle more in spring with seasonal colds because damp weather isn't good for them. And some wilt in the high heat of summer.

When you are looking at how you interact with the seasonal changes, you should be thinking in terms of your health and emotional wellness. You can also think about mental clarity. Often these three things are intertwined. I know lots of people who are vata types who have a naturally thin frame, fine hair, and thin skin, along with a variable appetite, and who are prone to insomnia. They get a little bit worn-out and scattered when they are imbalanced. Many of them struggle in late fall and early winter, the vata season of dry cold. In winter, their insomnia flares up. Their skin and tissues get dehydrated, which makes them prone to colds, and they forget to take care of themselves. They often feel anxious and overwhelmed on the shortest days of the year. They lack energy. For years, they have been telling themselves that the holiday season makes them crazy, but actually the holidays have very little to do with it. It's the season alone that affects them. If you live in an area where the cold and dry air is more intense, your body will react even more.

If this sounds familiar, you will want to take care of your body as the fall progresses. Make sure you get exercise every day, because this will warm your body and focus your mind. Drink herbal teas all day. Make sure you have adequate cover when you leave the house. This means covering your ears and sometimes you may want to put a little sesame oil on a Q-tip and massage your ear canal to keep it moist. You may want to keep a little bit of oil in an eyedropper and put a drop or two into each nasal passage at night. Oil massage is wonderful for your body in any season, but especially when the days are short. Massage your feet and legs with oil before you shower. You need more healthy fats in your diet than most people, so be sure to use oils or ghee in your cooking. Eat only warm foods—no raw salads or cold sandwiches or chilled yogurt in winter. These may seem like small measures, but they make an enormous difference.

If you are a strong sleeper with a slow metabolism who exercises only lightly, you probably have a kapha body type. You may gain weight much more easily than you lose it, and you may have some trouble with lethargy in the morning and congestion. For you, the spring will be the tough season with its high humidity and varying temperatures. You may find yourself getting sick during the kapha season, being more lethargic, and less able to get out of bed in the morning. You may gain weight in the late winter and early spring. This is another body type that reacts to the cold, but your body produces congestion in the moist, cold air. You will struggle in any climate with rainy, cold weather. To combat this weather, you need intense daily exercise that will give you warmth and flush fluids from your body. You will also benefit from a postworkout sauna that will further allow you to sweat out fluids. Unlike vata, you don't need fats in your diet, but you do need warm foods. Spicy foods are especially good for you because they open your nasal passages. Be strict about your exercise regimen and your diet during the late winter and spring months. Get up early and go to the gym and your whole day will unfold with more clarity.

Pitta types are those with lots of body heat. They have early morning insomnia, hearty appetites, and a strong need for exercise. These people may have trouble in the summer, when the temperatures run hot. When they are out of balance, they push themselves too hard physically and emotionally. If this is you, be careful as the summer becomes hot and muggy. A tropical climate is not for you because your body can't easily cool itself. You are more prone to headaches or even migraines in the bright light of summer. And inflammation will cause your old injuries to flare up. Avoid spicy foods during this time. Think of eating cooling things, such as raw fruits and vegetables and mild, savory spices. Cool yourself with swimming and cold showers. Even splashing your face or feet with cool water before bed will help you. Drink lots of water all day as this has a cooling effect, and stay away from alcohol. Dial back on your intense exercise in the summer months. Your body isn't as strong during this time of year and there's no benefit to pushing yourself with weight training or long runs.

Snowbird Syndrome

So what happens if you live in a cold climate and plan a lengthy stay in a tropical climate just as the days are growing short? From an Ayurvedic perspective, what happens is obvious. The body gets confused by the completely different light cycle, temperature, and food. It's as though you've plucked your body out of its natural hibernation cycle and plunged it into the productive season of gentle warmth, great activity, and a naturally high-carb diet. While your mood may be lifted by the warm sunshine and outdoor relaxation, your body may punish you in return with a nasty cold. Like it or not, our bodies are changing with the seasons, and you want to be aware of this before you travel so that you can prepare. Better yet, you want to travel when the seasons are beginning to change. That same winter stay after the solstice when the body is already beginning to adapt to the longer days would be much easier on your health.

Change Your Schedule, Change Your Life

I have a lot of patients who seek to avoid winter altogether by living half of the time in a warmer climate. When I lived in Hawaii, I worked with a couple who visited Hawaii each fall. They would arrive on the island of Kauai at Thanksgiving and stay until March or April, when—like clockwork—they would get very sick on their return to Toronto. Their annual arrival in Hawaii was no picnic, either, and they had trouble adjusting to the warm weather. The constant colds and exhaustion made them think they had some kind of allergies or depression, and they came to me hoping to cleanse their bodies of toxins. The detoxing did help, but they would still get sick when they went home. I asked them if it would be possible for them to change the dates of their visits. Could they manage to stay in Canada until the winter solstice and return before the spring equinox? It would mean staying in the cold for an extra month and returning home by the end of March when the weather was still cold. In this way, they could stay home and experience the beginning of winter and return home before it fully ended. Their bodies would experience the shortening of the days. After the winter solstice, the body is beginning to prepare for spring and they could leave for an extended stay elsewhere without shocking the system. If they returned before the days got too long they would be better able to adjust.

Once they adjusted their travel schedules to take these seasonal changes into account, they were able to spend most of the winter in a warmer climate without getting sick twice a year. By being aware of how the seasons affect your body, you can safely travel without shocking your system.

Because we live our lives largely indoors, it's tempting to think that the climate doesn't affect us, but it does. It changes everything about how our bodies function, from our moods to our physical health. By paying attention to the way your body reacts to the changing seasons, you can adjust your daily schedule to keep yourself healthy and grounded. Once you get used to doing this, it will become instinctual—the same way you pack up sweaters and coats in preparation for spring—and, in time, you'll adjust to the different rhythms of life throughout the year. You can take care of yourself in the season that is most challenging for you, and you can adjust your diet and exercise schedule to keep your radiant glow all year long. What's going on outside your window is also what's going on inside your body, and it's vital to notice how these changes are affecting you so that you can make the right changes to your diet and lifestyle to keep yourself in balance throughout the year.

11
The Seasons of Your Life

Marcus, a fifty-three-year-old real estate agent, came to see me shortly after he got married. He was beginning to feel sluggish and tired during the day, and unable to focus on work. In fact, it had gotten so bad that by noon each day, he was feeling that he needed a nap and couldn't meet with clients or make important phone calls in the afternoon. He was a stocky man, very muscular and tense. His new marriage was very fulfilling for him, but the rest of his life was falling apart, and he didn't know why. When we looked closer, though, we found that he had inadvertently adopted his new wife's schedule. Where before he had a rigorous morning exercise routine, a modest diet, and an early bedtime, he was now operating very differently. He slept in each morning because that's what his wife liked to do. They ate breakfast at eight and then both commuted. They would meet for lunch at a different restaurant every day. At night, she would start cooking dinner at seven or seven thirty, so they wouldn't sit down to eat until about nine p.m. And they would often have several glasses of wine before and during dinner. Then they would stay up together to watch TV until eleven p.m. before going to bed.

This would be a problematic schedule for anyone, but his wife wasn't having the same difficulties. She wasn't gaining as much weight as he was. She wasn't feeling as sluggish. She had yet to experience the true health problems with this kind of schedule. Why is that? Because she was twelve years younger, barely forty. And she had a different body type, one that naturally puts on less weight from an unhealthy schedule. It's likely that she would begin to have some problems, including insomnia and mental fatigue, if she continued with this schedule for too much longer, but it probably wouldn't manifest for a few years.

The body's circadian rhythm changes over the course of a lifetime. In the earliest months and years of life, your rhythm is a little bit erratic. As sleep-deprived new parents everywhere can attest, a baby doesn't strictly eat and sleep on a diurnal rhythm. But quickly in childhood, your body responds to these light and dark signals. In your young adult phase, the circadian rhythm is strongest. It can handle a little disruption and get back on track. At around age fifty, the traditional time of menopause and andropause, the natural circadian rhythm begins to weaken. In late middle age, people who have been relatively healthy can see sudden weight gain and new levels of fatigue and mental fogginess. You might have insomnia as a part of hormonal fluctuations. You might feel that these changes are a natural part of aging, but they don't have to be. By paying attention to your schedule at every phase of life, you can keep from gaining weight and feeling exhausted and transition gracefully and agelessly into the next decades of life.

I asked Marcus to bring his wife to his next appointment, because she was going to have to hear this information as well. Couples often create a tandem schedule as part of their daily communication. They want to do things together, and they could rearrange their schedules to better support their health.

They needed to make a few key changes. Marcus needed to get out of bed by six thirty in the morning and head out to a hot yoga class, where he could engage in the intense exercise his body needs first thing in the morning. At noon, they could still eat at a restaurant, but it would have to be healthy food. In the evenings, they would eat a light, early dinner, and his wife found this easier for her. It could be a salad or simple stir-fry, nothing elaborate. And they would forgo wine except as a weekend treat. In the evenings, they would turn off the television by eight thirty and spend time talking or reading before an early bedtime. By the end of one month, Marcus had lost weight and gained vitality, while his wife also lost a little weight and gained energy. She had better focus for her work and started to meditate in the morning while he was working out.

Part of paying attention to your body is paying attention to your stage of life. When you pass from one stage into the next, problems can emerge if you don't adjust your schedule to the body's changing needs.

The Ayurvedic Seasons of Life

According to Ayurveda, the body develops in three phases, which closely mimic the three dominant seasons of the year. Each season of life creates its own focus, both physically and emotionally, and we have to understand each phase in order to prepare the body for them. The kapha phase is a period of growth that lasts from birth to about twenty years of age. Pitta is characterized by stability and lasts from early adulthood (around age twenty) to about age forty-five or fifty. The final season of life, vata, can start in the late forties, but comes on gradually through your early fifties. Each of these stages offers emotional and physical challenges but also great opportunities for you to seize optimal health.

The Kapha Season

Kapha season is like springtime for your body. For the first twenty years, your body builds bones and tissues, and the circadian rhythm fluctuates wildly at times, trying to find a balance. Babies aren't born with a set sleep schedule, but they develop one quickly during the first months of life. Gradually, the body settles into a system in which the hormones, blood pressure, bowels, and other systems function on a diurnal schedule. Anyone with teenagers knows that they give up their regular sleep habits and become night owls. They are impossible to pry out of bed in the morning and sleep until noon on weekends. In fact, some researchers suggest that the real end of adolescence can be marked by the time when young adults give up trying to stay up so late. Teenagers' eating schedules, too, become erratic as they crave energy while their bodies are growing and maturing. When they get out of balance, teens can struggle in school and get inflammatory conditions, such as acne. They can adopt dietary habits that will be harder to shake as they become adults, which can lead to weight gain and depression in adulthood. This is a crucial time to introduce kids to healthy eating, a good night's sleep, and plenty of exercise. Their growing bodies demand a lot of fuel, and their muscles need to move in order to develop properly.

I often see patients who are still in their teen years struggling with school, friendships, and finding a sense of purpose. Though it may sound surprising, I can often trace these problems back to an unhealthy schedule, including late nights of doing homework (or texting while pretending to do homework), and eating unhealthy foods late in the day. Another culprit is little or no exercise, and a lack of natural light. Kids need natural light during these critical growing years. They need time outside and physical activity along with a good diet. Most parents enforce bedtimes to a certain point in childhood and then many of them give up. Make the effort to enforce a bedtime and time away from electronics. This will make an enormous difference in the quality of a teen's sleep and his or her grades at school.

Whenever parents bring a teenager to my clinic with social or academic difficulties, I immediately wonder if their daily health habits aren't supporting their growth. For teens, sleep, diet, and exercise habits are an overwhelming part of their identity. They define themselves by what their friends do, even if these habits sabotage their well-being. And they resist much of their parents' sensible advice about eating breakfast and getting to bed on time without electronic distractions. Ashley was brought to me by her parents. She attended a large public school and had lots of friends there, but her grades had suddenly dropped and her parents were worried. They'd been advised that she had ADHD and were considering medication. When she started her sophomore year, Ashley's parents pressured her to take advanced AP courses to help prepare her for college, and these classes had been a disaster. She began to bring home failing grades for the first time in her life, and she was becoming miserable. She complained of insomnia, and said that she never fell asleep before two in the morning.

She would lie in bed for hours tossing around, and then when she did sleep, she slept so deeply that she couldn't get up in the morning. She said that the insomnia was the result of all the pressure at school, but I had doubts about this as the only cause.

The rest of the day, Ashley had what many would consider to be a typical teenage schedule. She dragged herself out of bed as late as possible, skipped breakfast, and spent the morning in a fog. She ate a cafeteria lunch, which was more like a snack—lots of bread and starches that made her sleepy—and after school she sat at home dreading her homework. She ate a family meal at seven thirty or eight p.m., late enough for her parents to get home from work and prepare food. After dinner, she reluctantly did homework until ten p.m. and texted friends or watched YouTube videos until eleven or eleven thirty. And then she would lie awake, waiting to fall asleep, while dreading the next day in school. This schedule is common among kids. Some of them are out at night at the mall or a coffee shop until ten thirty or eleven, and then they come home and wonder why they can't sleep. Here is what I tell parents who are trying to set a healthy schedule for their kids:

Make social time exercise time. It's discouraging for me to see kids absorbed in their screens or sitting bored in front of the TV while the sun is shining outside. For many teens, a trip to the mall is their idea of socialization and exercise. I advise time outside doing something athletic, even if it's just a walk after school. In their teenage years, kids would do well to join a team or do some fitness classes. Martial arts is a great choice. Fortunately, Ashley loved to play tennis, and so I told her to go to a local park or a tennis club every day just to hit some balls around. She found a couple of friends who would do this with her.

She would do this for about an hour or ninety minutes every day after school. It refreshed her. Exercise gives children and teenagers a break from long hours at school. Some teenagers need quite a lot of exercise to hit the reset button on their moods and frustrations. Some need it first thing to help them shake off the morning lethargy. When you wake up with energy, you are better able to sleep later in the evening. Ashley had trouble waking up in the morning, so I asked her to use the treadmill in the basement for about twenty minutes. She had to get up at six, do some work on the treadmill, and have a shower. But by the time she went to school, she was wide awake and focused.

No more cereal. Cold cereal is just a glorified snack food. So are granola bars and those little containers of sweetened yogurt. These choices deliver a shot of energy that's gone in less than an hour. Most kids need whole, cooked grains or some protein. You can put oatmeal in a Crock-Pot overnight and have it ready first thing if you like. Add some nuts and fruit and the kids are good to go for the first three or four hours of school. Once Ashley got her morning routine set up for exercise, a shower, and a bowl of oatmeal, her morning classes became much easier.

Break the sugar and flour habits now. A lot of kids and teenagers come home hungry after school, and it's easy for them to settle into snacking on baked goods and sugars. Teenagers sometimes turn to coffee-based drinks or sodas with caffeine after school to deal with the mental exhaustion. This is a difficult thing for parents to manage because they may be working or distracted, and sometimes it's hard to lay down the law when kids are restless and cranky. Ashley was doing this, too, out of boredom. And she was eating crackers and drinking soda all afternoon, then bingeing on bread at dinner instead of whole foods. In the first phase of life, we don't worry too much about calories because kids need so much energy to function and grow. But in the afternoon and evening, you want them eating foods that will nourish them, otherwise they stay irritated and distracted, even when their bellies are full. Because this is the phase of life when the body and brain are growing, kids and teens need good nutrition to keep them fueled. I had to tell Ashley no more crackers and no bread with dinner. No more sugary drinks. Her parents could keep Tupperware filled with sliced carrots, celery, and cherry tomatoes that she could enjoy with a little bit of dressing. They should keep lots of fruit on hand. These would be her go-to snacks, along with plenty of water.

Institute a screen curfew. Now is the time to let kids and teens know that they need a break from electronic stimulation at night. Some parents confiscate smartphones at nine p.m. Others restrict the television. There should be a set curfew on electronics. In addition, some kids need extra help winding down. They may need a shower at night so that the warm water can help settle them. Others need a little warm milk in the evening while they read. Ashley tried oil massage. As a naturally light sleeper, she needed to massage her feet and hands with oil, and that helped her discharge the daily stress. After that, she would read and shut out the lights at ten or ten fifteen. Within a week of setting up this schedule, she was able to fall asleep much more easily. After a month, her grades and mood had rebounded. This is the miracle of the kapha phase of life. Small changes can make a big difference very quickly.

The Pitta Season

Pitta season is the summer season for your body. You have a lot of drive and energy in these years and a real desire to get things done. Between the ages of twenty and fifty, your body has matured and your circadian rhythm has hit its stride. In your twenties, your digestion is as strong as it will ever be and your metabolism is on fire. Some people can defy the body's needs by staying up late, or eating junk food, or refusing to exercise, though few people can get away with all three. Still, this golden period of seemingly effortless health begins to slip away within a decade, if it ever truly existed. Some body types are more likely to gain weight even during these years, while others are prone to insomnia or workaholism. This is the time of life when your body will be the most forgiving of disruption because the circadian rhythm is strongest. But you do still need a set schedule to keep your body fine-tuned so you have more focus at work and a greater sense of purpose in your life. And staying on track now will set you up to age gracefully in the decades ahead.

The earliest part of this phase is when people are more prone to obsession. They throw themselves into a particular passion or work project and forget to take care of their bodies. Sometimes their bodies rebel. This is what happened to Jamal, a twenty-six-year-old software developer. You can probably guess his schedule: long hours, plenty of takeout, and barely any exercise. His kitchen at home was a sterile environment. The oven had never been turned on. No piece of fresh produce had ever touched the inside of his refrigerator. Instead, it was stocked with condiments, energy drinks, and beer. He was cheerful about his ability to stay up all night completing a project, and he never thought twice about devouring a bag of chips in one sitting. And yet he was noticing that his hair was thinning on top. The double chin didn't make him feel attractive. He was waking up at four every morning, unable to get back to sleep. And he felt a stiffness in his joints, a lethargy that he couldn't shake.

His physician would have drawn blood and then told him to change his diet, and would probably have warned him that he was on the path to metabolic trouble. But this kind of pitta imbalance has deeper roots than simply diet. If you are in this phase of life, you have to look at your health habits holistically. Your body operates on a central rhythm, and all of your daily habits affect that rhythm. When you binge on work or food or entertainment or a relationship, you throw the whole body out of balance at a time in your life when you want to support your youthful vitality rather than sabotage it. Supporting healthy habits now will pay dividends for decades to come. Here are some guidelines to keep yourself in balance:

Choose your friends carefully. When you have diet and sleep problems in your twenties and thirties, you should look at your friends. Look at the waistlines around you, because you are probably eating what your friends eat. You are exercising when and if they exercise, and you are setting your sleeping schedule around your social schedule or your work life. As a teenager, you had the moderating influence of your parents, but in college and beyond, you are under the sway of the people you spend time with. This includes romantic partners. You are also beholden to the culture of your workplace. Be careful here. Most people think of themselves as individuals, but few people can really resist the temptations served up by friends and coworkers. In Jamal's case, he was in a job that prioritized long hours, binge eating, and no exercise. On the weekends, he was hanging out with the same people from work. So his weekends were spent at bars, drinking beer and eating pizza. He needed to join a gym and introduce himself to people who put fitness first, people who already prioritized healthy diets. Eventually, he joined a group of coworkers who loved competitive biking, and this became his passion as well. I often tell people in their early twenties that they need to find friends who are out of their league. Hang out with people who are already doing what you want to do with your life and who are already living healthily. Do this and you will much more easily change your habits.

Balance your exercise with stretching. Some people do get involved with fitness in their pitta phase, because it's a natural fit. Your body is strongest now, and you can push yourself without the same fear of exhaustion or injury that you will face later. One big mistake is to sign up to run a 5K or engage in triathlon training without making the time to stretch and breathe. You are on the go all the time in this phase of life, always squeezing in more of everything. That's fine, but taking the time to stretch will keep your muscles supple for the second half of this phase. If you get injured in your early thirties from doing too much too fast, you'll find your fitness level slipping away. If this happens, your motivation for exercise will surely fly out the door as well. Jamal found a power yoga class specifically for runners, and he loved it. Any type of yoga or stretching program would have worked just as well. On those long workdays, he took some time to walk after lunch, to get some sunshine and stretch his legs. By doing this, he was taking care of the body he will need for decades to come.

Pay attention to your bedtime. The pitta phase is prime time for insomnia as work and family schedules get more intense. It's easy to say that you can sleep more later or on the weekends, but being vigilant about your bedtime now will carry enormous benefits for you at work and in your relationships. Yes, there will be months at a time when a baby's schedule will interfere with this. Yes, you will have work deadlines and travel that can't wait. This is all normal and temporary. During those stressful times, you can turn to meditation; even fifteen minutes of mindful rest each day will keep you mentally sharp. When the crisis is over, be sure to check in with yourself to make sure that you have downtime in the evening and can turn out the lights by ten p.m.

Detox regularly. This can mean a number of things. Early in pitta season, you want to be learning to cook healthy food at home so you can minimize your takeout menus. Relying too much on food that comes from a delivery person or from a box in the freezer is setting you up to age prematurely. Jamal had to learn to shop for ingredients for simple salads, and he needed a Crock-Pot to make soups and stews. Taking the time to do this put him in touch with his body. He also needed seasonal cleanses to reset his metabolism. Some people need to do the seasonal cleanse with the change of every season, while others can do it once a year at the time when they struggle the most. Even if you don't do the full castor oil cleanse, you can use the detox diet for a week or so as the seasons change to help you take stock of your diet and keep yourself on track.

At the end of three months of making simple changes in his diet and his exercise routine, and getting good sleep, Jamal felt like a new man. He said he felt healthier than he ever had, even in high school and college.

The Vata Season

This late autumn season of life creeps in slowly. Some people notice its physical effects in their late forties, while for others, it's felt in the midfifties. You may first notice that your eyesight changes and you need reading glasses. Your metabolism slows and it's harder to lose weight or maintain your weight. You may experience metabolic issues as your blood sugar level and blood pressure rise. Your skin may become drier, and your muscles don't hold the same firm tone that they once did. Your physician might say that these are natural and expected changes that come at this time, and they are. But you can delay their arrival (and maintain a high quality of life once they do) by paying attention to your body's natural rhythms. This is the time when your schedule can either slow the aging process or accelerate it. Your circadian rhythm is weakening as well, which means that an erratic eating, sleeping, and exercise schedule further disrupts your body. A consistent schedule reinforces your body's systems while you age.

When patients in this stage come to my office, I sometimes have them get on a special scale that measures several factors in their body composition, including weight, body fat, BMI, skeletal muscle, and resting metabolism. Using these measurements, the scale then calculates their body's approximate age. People are sometimes shocked to learn that their body measurements indicate that their biological age is eight to ten years older than their chronological age. This is the wake-up call they need. I know my patients can feel fifty when they are actually seventy years old, but most people are going in the opposite direction. Once they get on my scale, patients sometimes discover that their approximate body age is sixty-five, when they are, in fact, in their midfifties. They see clearly that the extra pounds, the higher than normal blood pressure, and their daily lethargy are not normal parts of the aging process. Instead, these are alarm signals that your body is aging faster than it needs to. After a few months on a good schedule, they get back on the scale and watch the years roll back. You can do the same thing.

Connect to your body. The vata season is a natural time for reflection. You look around at the life you have built and the connections you have made and feel good about everything you've done so far. You aren't giving up on setting more goals. Far from it. But this is a time to renew your purpose and make meaningful choices about how to spend your time. It's important to extend these reflective thoughts to your body. People who age slowly are people who pay attention to the body's signals instead of their moment-to-moment desires. That means resting when you are exhausted, getting more exercise when you feel lethargic. It also means taking a hard look at your diet and other habits that are causing you harm. A regular meditative practice can make all the difference here. When you focus on your breath and your body for just fifteen minutes a day, you are actively listening to the signals it is sending you about which foods are healthy and which ones are a drag on your system. You will also need fewer hours of sleep at night. That doesn't mean you should stay up late, but that you can get up earlier in the morning.

Maintain your balance. This means a couple of things. The most obvious is to do physical exercises that require balance and flexibility. Many yoga poses improve balance, as do isometric exercises and cardio exercises such as jumping rope. Flexibility is part of balance, so you must stretch regularly. The ability to touch your toes or stand on one foot might not seem like a big deal on your fiftieth birthday, but these are skills you want to preserve now. You can't continue to hike and bike and play tennis or lift weights if you lose either balance or flexibility. But balance here also means maintaining your emotional equilibrium in general. After age fifty, you need to check in with your body more often, to know when you feel rattled or anxious, to know how travel affects you, and to cultivate the practice of restoring physical balance after hectic or stressful times. By this point in your life, you have accumulated a lot of wisdom about how you move in the world and what pulls you off center. Use that wisdom to rest when you need to and to alleviate stress with meditation and reflection.

Reduce inflammation. A low level of whole body inflammation is thought to contribute to all kinds of aging-related illnesses, including dementia, heart disease, and cancer. If you are taking over-the-counter pain medication daily, you probably have some form of this whole-body inflammation that is settling into your joints. In order to reduce this inflammation, you need to keep your weight down, eat cleanly, and regulate your levels of stress. Your body will also need more time to recover from intense exercise. Plan enough rest between workouts to keep injury at bay and to keep yourself from needing pain relievers. When you need additional support reducing inflammation, you can take turmeric supplements. Turmeric can help you lessen your reliance on ibuprofen, and is certainly healthier for your stomach.

Prevent dehydration. The tissues in your body will be less likely to retain fluids as you age. You will notice this primarily as dry skin, but your whole body will be affected. It may be a good time to reduce your daily intake of coffee and tea, which can further dry out your system. Build a diet that includes healthy fats to lubricate joints and ligaments, and try oil massage as an excellent way to get healthy oils into your skin and body. Two of my patients had moved to Arizona when they were in their forties and loved it because of the hot, dry weather. But by the time they had turned seventy, they hated the heat and told me they couldn't even bring themselves to look out the window at the barren landscape. It was clear to me that the dehydration of the climate was affecting them in a way it hadn't when they were younger. Eventually, they moved to Florida, where there is plenty of heat, but cool mornings and lots of humidity. Their moods and their health improved dramatically. Dehydration can sneak up on you in many ways. That's why it's so important to understand the needs of the body.

Simplify your diet. If you are still eating like you're in the pitta, or even kapha, season of life, you will bear the consequences now. Many people notice heartburn, lactose intolerance, sensitivity to certain foods, or sudden weight gain at this time. These are signals from your body that your diet is no longer nourishing you the way it should. You may feel more lethargic after a big meal. In some cases, this heavy feeling may last for days after an indulgence. I tell people that a man of twenty can eat a large steak and feel the effects for just a couple of hours. A man of forty-five will eat a few slices of that same steak and feel the effects for a couple of days. A man of seventy can eat just a few bites and will feel the effects for two weeks. The digestive fire is naturally growing weaker, which means you need fewer calories and less meat to stay robust. You can eat smaller, lighter meals and still feel satisfied. It's important to view dairy, flour, and sugar with suspicion. This is the time to practice the seasonal cleanses four times a year and to practice gentle fasting in which you give up dinner once a week. You are less hungry at this stage of life, and your diet can reflect this with a reduction in meal size.

By doing these things, you will be able to attend to your vata body's changing needs and—as the circadian rhythm becomes weaker—you can remain strong in your healthy schedule. I have patients who look youthful and feel vibrant past the age of seventy. They are busy and active, still running businesses, still enjoying busy social calendars. This is a true reversal of aging.

We are always encouraged to look at age as a defining characteristic of identity—an approaching birthday is either something to celebrate or something to dread. But instead of focusing on your biological age, you can and *should* think in terms of your season of life and what you can do to support the body's natural maturation. Remember, your wellness is not solely determined by your age, and there is much you can do to delay or even reverse the negative aspects of aging. No matter how old you are, you need to take time to check in with your body and asses its health. If you do this, you are accumulating wisdom about your body's needs, about your goals, and about your purpose in life as each birthday goes by. And that wisdom is yours to keep.

12

How to Build the Perfect Day

The decision to live differently is a challenge for everyone: Corporate CEOs complain that they can't stop working sixteen-hour days and traveling all over the globe or their careers will collapse. Elite athletes tell me they can't stop overtraining or they can't stop the continual search for sponsors because they won't be themselves if they don't keep doing everything they've done in the past. All kinds of people tell me that they don't have time to exercise before their morning commute, that they can't stop working through their lunch hours, or that they won't give up on late-night TV. Habits become part of your identity, and it's scary to try something new. You have to remember that you didn't develop your unhealthy schedule and poor dietary choices in one go. Your daily schedule has evolved over years—maybe decades. Some of your habits have emerged as a coping strategy for stress, as a way to find comfort in a chaotic life. You can seek a deeper level of security by challenging these habits one by one.

If you're like most of my patients, you'll have a list of reasons to keep your life exactly as it is, even if you are unhappy. You want to lose weight, but your mind is telling you to cling to the old diet because it offers some comfort. You want to get more sleep, but you believe that you deserve a couple of hours to relax at the end of the day. Maybe you are intrigued by the idea that exercise will give you more energy, but your mind interferes again, saying there is no time for that. In reality, your mind is telling you that you don't deserve better health, but you do. No matter how you have lived in the past, you deserve to feel better now. Deciding to change is the hardest part.

By now it should be relatively easy to imagine the level of good health you can achieve if you will let go of unhealthy habits and get in tune with your body, but knowing what's going on in your body doesn't mean change will be easy and convenient. This was the case with Rowena. A little over five feet tall and 237 pounds, clinically depressed, and steadily gaining weight, her schedule was literally killing her. Because she worked the night shift, she would get home from work at eight a.m. and eat toaster pastries for breakfast while watching TV. Then she would go shopping in the morning to get out of the house. At the mall, she would eat lunch at the food court—a different fast-food place every day. In the afternoon, she napped off and on until about eight p.m. Then she would heat up a frozen dinner and go off to work with a large thermos of coffee to help keep her awake all night.

She was running out of time and she knew it. Still, she gave me all of the reasons why she couldn't change anything. Like most people, she was clinging to her old diet and habits because she thought they were giving her the only comfort she could count on, even though her state of health was making her miserable. Even after she understood that she could use her body's rhythm to promote weight loss and increase her energy, she was afraid to change anything. She seemed overwhelmed by the idea of a new schedule. It would mean a lot of daily changes and a whole new diet. It was a lot to remember. I wasn't sure I would ever see her again.

A little over two weeks later, she came back. She had lost eight pounds, something she had never done on any other diet. She gave up going to the mall and started walking on the beach every morning, getting her exercise. She had decided to retire from her job and do some part-time work during the day, so she could sleep at night. It was a big decision, but her new schedule gave her a new sense of energy. More important, it gave her some control over her life. For the first time in years, she felt she had the power over food cravings.

I saw this as the beginning of real change for Rowena. But what's really important is that she had started to see this possibility, too. She started to eat better, meal by meal. She decided to get some exercise day after day. She decided to make new choices about how to spend her time at work. And her life was already expanding in ways she couldn't have imagined. That's the point. I include this example here because deciding to change your habits day by day can add up to an enormous shift in perspective. Obviously, some changes are easier than others, but changing your life is a process, one that you want to start as soon as possible.

The Perfect Day

Imagine a perfectly scheduled day, a day in which you know how you are going to take care of your body as soon as you open your eyes. You are going to exercise without needing extra motivation and without overdoing it. You are going to eat well without a struggle, knowing that you will feel light and energized all day without being hungry. You know when you are going to meditate, to center yourself, so that you can deal with any kind of stress. You know already that you are going to be able to face challenges at work, and at the end of the day, you are going to feel pleasantly tired so that when you shut off the bedside lamp you are ready to lie quietly until sleep comes.

All of this is possible once you align your body's systems with its natural circadian rhythm. This is the real goal, the ability to eat, sleep, and exercise at the right time each day so that the habits don't require a lot of extra effort. Wellness and self-care become an effortless, intuitive way of living. For many of the people I've worked with, having a good body schedule in place actually opens up extra time in the day to spend with family, to engage in hobbies, and to find new relationships.

If you are wondering how to get to this mystical place, the answer is that you do it in stages. You are going to build these habits bit by bit, knowing that the first steps may require you to expend some effort, but you will very quickly feel lighter and more alive. Many of the patients whose stories I've shared in this book are people who needed a lot of concentrated help to get their lives back on schedule. But once they started, they found each successive step easier. By the time they'd spent a month trying to change their bedtimes and eating schedules, they found so many benefits. More free time, more creativity to focus on their personal goals. They looked radiant and felt more energetic than they had in years. Intuitive, effortless wellness is within your reach, too.

The question is how to get to that place. Now that you know what your body needs, you can work toward building a new schedule in stages. At first, you will be setting up a schedule that supports the body's natural circadian rhythm. This is the phase that will require the most preparation and creativity on your part to live a little differently, but it will all pay off in phase two, when the schedule becomes easier to stick to. In phase two, people start to notice weight loss, an increase in mental clarity, and dramatically improved health. In phase three, we are putting together all of the concepts in this book so that you are easily nourishing your body, your mind, and your spiritual health.

Preparing for Change

In every meeting with a new patient, there is a moment when I lay out their new schedule and ask directly, *Can you do this?* If they hesitate, I then ask, *What is your biggest resistance?* It's important to acknowledge which part of this schedule seems most problematic for you. It's different for everyone. Some people say that they don't think they can stay away from work emails at night. Others say that they don't want to eat a lighter meal at the end of the day because it's their one chance to relax with a lot of food and wine after feeling stressed at work. A few object to exercise in the morning because their mornings are already so hectic. But once you know what your resistance is, you can look for a creative way to solve it. Be mindful of the potential challenges ahead, and get organized before you start. Figure out which parts of this schedule will be most challenging for you, and plan to use some of these tips and tricks to ease your transformation:

If morning exercise is the problem . . . Put oatmeal into a Crock-Pot overnight to make time for breakfast. Set your alarm five minutes earlier each day. In just four days, you will have enough time for that morning walk. Lay out your workout clothes and your workday clothes the night before.

If a consistent lunch sounds intimidating . . . Pack a substantial noon meal to take to work, and remember that a bigger lunch will mean less elaborate cooking at dinner. Set an alarm on your phone to remind you to eat lunch on time.

If you worry about how to eat less at dinner . . . Look for some easy but healthy dinner options you can rely on during the first phase so you can finish your evening meal earlier. You can even make them ahead, if you know the week will be hectic at work. Create a standard grocery list for yourself that contains lots of vegetables and healthy grains.

If the thought of giving up electronics in the evening makes you anxious . . . Plan evening activities now that don't include the TV or computer. Think about whether you really get texts or emails after nine p.m. that truly must be answered.

By applying a little creativity and planning to your new schedule before you start, you can avoid falling into your old habits after the first few days. By the end of the first week, you will have noticed enough physical and emotional improvements to keep you going.

Phase One: Balance

For the first week to ten days, focus on reconnecting to your body's natural circadian rhythm. That means getting up on time, getting to sleep on time, and eating your largest meal at noon with a smaller meal in the evening. You will also introduce a little bit of exercise first thing in the morning. If you can get to a gym, great. If not, try to go outside for a walk or do some exercises at home. Many of my patients have an old treadmill that has been idle for several years. This is the time to get it out and do some brisk walking or a slow jog, as little as ten minutes or as much as twenty minutes. Remember, you aren't aiming for burning calories or fat. What you want is the lungs breathing deeply, and a light sweat. This is your opportunity to foster prana, and to remind your body that the day has started. If you love to exercise intensely, go ahead and move your full workout to the morning.

Work on getting meal timing right. At first, you will likely need to set an alarm on your watch or phone to remind you to eat breakfast, lunch, and dinner on time, but by the end of this phase your hunger will be a natural prompt, and the timing will feel more comfortable. You will get used to the routine of the morning, getting up and moving, followed by eating. Then your morning commute.

The evening ritual will also become natural. The first night you may feel restless without the TV and the usual computer distractions. Find some personal projects that you have been putting off. Pamper yourself with personal grooming or a nice bath. Savor this time and it will quickly become the best part of the day.

It's also important to find time for a little bit of mindfulness. You can take your pulse a few times a day. Just sit and listen to it. Or you can sit in bed right before sleep and close your eyes and let your mind drift while you breathe deeply. You can also write in a journal at the end of the day. Any one of these things will start you on the path of reconnecting to your body.

In phase one, your day will look like this:

6:00 a.m.—Get out of bed. It's important to start the day on time.

6:15 a.m.—Wake up your body with twenty minutes of brisk exercise. At this point, you are just getting used to the idea of getting up and getting moving first thing.

7:00 a.m.—Eat a modest breakfast. Choose between oatmeal, or a smoothie, or eggs with steamed veggies. Finish your meal by eight thirty and plan to eat nothing—not even coffee or tea—until noon.

12:00 p.m.—Eat a substantial lunch near the noon hour, but be sure to eat at the same time every day. Set the alarm on your phone if you need a reminder. This should be the largest meal of your day. Remember to check your pulse before and after you eat.

12:30 p.m.—Try to take a short walk after lunch or spend some time outside to get natural light. This isn't exercise, per se, but rather gentle movement that will aid your digestion. Sit by a window if you can't get outside.

1:00 p.m.—Refill your water bottle and make a cup of herbal tea. Set your intention to avoid snacks in the afternoon.

6:30 p.m.—Eat a light dinner. Focus on vegetables and a small portion of protein, if any. Avoid rice, bread, simple sugars. Look to consume about five hundred calories for this meal. If you are still at work, you can eat a light dinner there. If your dinner is delayed because of a commute, remember to reduce your dinner portion accordingly. The later you eat, the smaller the meal should be. Last call for any food is eight p.m. You need at least two hours after dinner to prepare for sleep.

8:30 p.m.—Send your last texts and emails of the evening, and shut off your computer and the TV. (DVR your favorite shows for tomorrow if you need to.) This begins your evening routine of reading, meditating, taking a bath, journaling, or whatever else you want to do. This is your time to relax and be calm.

10:30 p.m.—Turn out the lights. Even if you have trouble falling asleep right away the first couple of nights, you can breathe deeply and relax as you wait for sleep.

Did You Remember To:

- Take your pulse several times a day? You aren't listening for any pattern. Just appreciate your heartbeat and notice your body. Do this when you wake up and before you turn out the lights.
- Close your eyes and take some deep breaths when you felt stressed? Discharging stress makes it easier to resist reaching out for distractions, including junk food or your phone. It calms your mind and resets the body.
- Ask how your body feels after each meal? If it helps, make a note in your food journal. You want to strengthen the mind-body connection.

Herbs for Phase One

People often pay more attention to their eating habits than to their bowel movements. But you can't just pay attention to what goes into your system. You need to be producing something on the other end. Your intake and your output are both measures of your good health. Triphala and trikatu are two herbs that can help this process. They are easily available online through the Chopra Center or Banyan Botanicals, or from my own website.

Triphala. This is a supplement that has been used for thousands of years in Ayurveda. It means "the three fruits" because it is a mixture of the powders of three berries. While it is known as a mild laxative, it really encourages detoxification of the bowels and the body. It helps regulate blood sugar while relieving constipation. Taking about one thousand milligrams every night helps regulate your bowel movements so that you can go first thing in the morning and prepare your digestion for more nutrients.

Trikatu. This supplement helps improve your digestion and allows you to absorb more nutrients from your food. This formula contains three spices and is often called "the three peppers." Where triphala helps digestion in the lower intestines, trikatu helps fire up digestion in the stomach and upper intestines. Take five hundred milligrams after lunch and dinner.

Stay in phase one for seven to ten days, until you feel comfortable scheduling your day around sleeping and eating with a little bit of exercise first thing. You will notice that you start to feel sleepy at your new bedtime. You may also begin to notice that you aren't as hungry for your evening meal. You should also be waking up on time more naturally. Some people say their bodies feel lighter right away and their bowels change. You may begin to have regular morning bowel movements within the first week, even if you never had them before.

Phase Two: Healing

In this phase, we are going to fine-tune your schedule and fire up your digestion and deepen your sleep at night. We are looking for an effortless schedule, something that is healthy and feels natural. This is the phase when you will start to see weight loss, and it is also a time to pay attention to food cravings so that you don't get derailed. Each time you go to the grocery store, you want to choose freshly made foods and turn away from prepackaged foods. We are also adding in some mindfulness techniques so that you make the connection between your schedule and how you feel during the day. If you can find several times during the day to get up from your desk and go outside for a short walk, you will be giving yourself more natural light and getting some movement into your day. In this phase, you will see the true health benefits of putting your body's needs first.

You should notice that you are starting to get hungrier at the right time. You may not yet feel hungry for breakfast, but it will start to feel good to eat something in the morning. This first meal of the day makes it so much easier to resist coffee or snacks in the morning. By lunch you will have a real appetite, so don't be afraid to satisfy it. Remember, this is your most substantial meal of the day. In this phase, we are looking to deepen this practice by choosing healthier meal options. That means foods with less heavy meat, less bread, and more vegetables. You can eat plenty of food at lunch; make it healthy. After lunch, you may have three or four hours in which you feel full and snacks are completely unappetizing, particularly if you are drinking lots of water. You won't need afternoon coffee anymore. Your body will feel lighter between meals, and you will no longer feel ravenous in the evening. This will make it easier to avoid breads and sweets in the evening, which is a must. Your evening meal will be a light soup or salad or a simple stir-fry. Avoid foods that come from a box or from the freezer.

By now your evening routine is set and falling asleep is easier than it has ever been. You may be drowsy by ten p.m., and that's fine. You can sleep earlier because that will make it even easier to get up the next morning.

Because you are sleeping more regular hours, your sleep will become deeper and your bowels will feel better. The morning bowel movement will become a regular event, and you will feel lighter afterward. In fact, your whole body will feel lighter, and your bathroom scale should agree with this feeling.

In phase two, you will use the structure established in phase one, but build on it. In this phase, your day looks like this:

6:00 a.m.—Get up and out of bed. Drink a cup of hot water first thing to help start your digestion. You can even fill a thermos with hot water and put it by your bed the night before, if you like. The hot water gets your digestion and bowels going right away.

6:15 a.m.—Exercise. By now you want to be increasing your exercise intensity. Add some interval training to your morning walk by jogging for one minute, followed by walking for two. You can also add isometric exercises, or yoga poses. Don't overdo it! As you continue to exercise, keep pushing your body to get that good, deep breathing.

7:00 a.m.—Breakfast. Keep it light and healthy.

8:30 a.m.—By now you are used to finishing your coffee or tea and setting your intention not to snack. Drink only water or herbal tea until lunch.

10:00 a.m.—At the midmorning point, get up from your desk and move around. You can do this any time you start to feel sleepy or lethargic at your desk. You want to incorporate movement into your day. Get outside if you can, or take some stairs. Deep breaths build energy.

12:00 p.m.—At this point, you know to eat your largest meal at noon. By now your appetite will naturally match your mealtime. This is the first sign that your body is syncing with your circadian rhythm. Your body is hungry at the time of day when it most needs calories. Even though you now have a bigger appetite at noon, you still need to eat healthy, vegetable-based options. Avoid heavy, flour-based foods.

12:30 p.m.—Take a short walk outside, which will aid digestion and give you some natural light.

1:00 p.m.—Set your intention to avoid snacks in the afternoon. By giving your body a break from food for five or six hours, you are allowing your digestive tract to rest and reset its circadian rhythm. This is an important step in healing the brain-gut axis. It also allows your body to stoke a gentle hunger for a light evening meal. Plan to drink two to three glasses of water or cups of herbal tea in the afternoon. Make sure you are hydrating your body. This will give you a healthy glow.

3:00 p.m.—Get up and move around again. Do a little bit of stretching at your desk or walk around the block. This will give you more energy to power through the workday and give you more natural light.

6:30 p.m.—Eat your light dinner. By now you will be used to eating a lighter evening meal and finding it satisfying, even though it's not the large, heavy meal you used to eat. If you eat now, you will have twelve hours of natural fasting before breakfast, which allows your digestive tract to repair itself. This will make your breakfast more satisfying and help your bowels to recover overnight.

8:30 p.m.—Joyfully turn off your computer and phone, knowing that the best part of the day has started. By now you should have a set routine for the evening. It may include time with family, making plans for the future, reading, or reflecting. This is your time to be present with yourself and your loved ones, and to reflect on your goals.

10:00 to 10:30 p.m.—Lights out for a deep sleep.

Did You Remember To:

- Look at your stool first thing in the morning? How have your bowel movements changed during the days when you have been keeping a healthy schedule? With more fiber and water in your diet, your colon may be thanking you each day.

- Look at your tongue each morning? You should notice that the coating on your tongue has changed color. It has become lighter as the toxins leave your body.
- Take your pulse after each meal? This will give you a sense of how your body responds to the food you eat. You are listening to your heartbeat, but you are also listening to your body's response to different meals.

Herbs for Phase Two

As you refine your eating and sleeping routine, you will be accumulating fewer toxins. Your channels are more open and your energy is flowing. That means you are ready to begin burning more fat. Remember, these herbs aren't a replacement for building a strong routine and listening to the wisdom of your body type; however, they can give you a boost.

Guggul. This herb helps your body burn unwanted fat. It has also been shown to reduce cholesterol levels. It supports heart health and thyroid function. Guggul also helps purge congestion and mucus for people with a slower metabolism. The trick is that you need to have some healthy habits in place in order for this to do its work. You can take this three times a day, once after each meal.

Amalaki. This herb is often called Indian gooseberry. While many people think of it as a general tonic for the system, it has a powerful ability to reduce inflammation. It reduces heat and hyperacidity in the body. People who are feeling tired, fatigued, or having signs of inflammation from exercise or stress can benefit from this herb. Use this to cool down the digestion and reduce food cravings or heartburn. So this is a great herb to use when you are increasing your level of exercise and trying to make better food choices. (Note: if you are taking triphala, you are already taking amalaki as it's one of the berries included in that formulation.)

You should stay in the healing phase for ten to fourteen days, maybe a little more. Some patients stay here for up to three months while they fine-tune their schedule. This is the time when you will see real changes in your ability to wake up first thing in the morning, have more clarity at work, and notice physical changes in your fitness level. You are also gradually purifying your diet by paying attention to your moods after you eat, so refined flours and sugars won't be as appealing. The goal is to build your schedule around the body's needs rather than your work and old habits. You may begin to notice that if something disrupts your schedule, you feel the difference right away. If there is a nighttime work emergency or something, you feel the difference in your ability to sleep that night and your mental clarity the next day.

Phase Three: Transformation

By now you feel comfortable—happy even—navigating your new schedule. You've lost weight, gained focus, and are able to sleep well at night. So now what? If you've ever tried a new diet and exercise routine only to abandon it after a few weeks or months, you know what happens next: the maintenance phase, which usually tolls the death knell of your diet.

In phase one, you got by on adrenaline or excitement. In phase two, you saw results and those results brought their own challenges. But maintenance is boring and challenging, and it brings fewer immediate rewards to keep you motivated. So, in phase three, we want to combat boredom with mindfulness and deepen your practice by adding true meditation.

Still flush with the success you've encountered in the first two phases, your mind is going to play tricks, telling you that you can give up exercise some days if you are still doing everything else. Or it will tell you that you can answer a couple of emails late at night. Or it will tell you that feasting at night won't hurt anything. I know people go through this because they send me emails telling me that they've been good for a few months and asking if they can have pizza as a reward? No. How about a bagel? No. How about a bite of a bagel? Just one bite? Now, one bite of a bagel won't derail your whole diet; however, what most people are actually reaching for isn't a treat, it's stress management.

Cultivating mindfulness throughout the day will keep you on track with your new schedule and give you a new perspective. Strengthening your awareness of both mind and body will help you move away from thinking your diet is something punitive and begin to see it as something nourishing. If you believe that your morning exercise is a punishment for being out of shape, you have set the expectation that you can quit once you've suffered enough or gotten fit enough. This lifestyle has nothing to do with suffering. It's about giving your body what it needs so you can get more out of life.

Moving Your Set Point

I've worked with thousands of people to bring them out of the cycle of unhealthy habits and into glowing good health. I've noticed that the major barrier to change is emotional. People often develop poor habits because they crave a comfort that was denied to them early in life. Early emotional trauma creates a kind of psychological set point in which you decide that you don't deserve love and fulfillment. This is the lens through which you view every possible choice. When you have a low set point, you reach for what comfort you can find in junk food or other distractions. These habits create further blockages in the body. An unhealthy schedule blocks the body's channels and lowers your prana, your vital energy. These blockages reinforce that feeling from childhood that you don't deserve better, that you don't deserve to feel fully alive. But you do.

What I am proposing is a total shake-up of your old routine and your old habits. You may continue to resist it, even after you have started. Your mind will be telling you that you shouldn't have to stick to this diet, or do this early morning exercise, or get to bed on time, even if your body feels better than it ever has. When you move out of phase two and into phase three of this program, you will need a meditation practice to help you change your emotional set point. You can create a new lens through which you view the world. Soon, it will feel natural to nurture yourself with healthier food, an early but consistent bedtime, and daily movement. You will also feel less stressed and better able to build stronger relationships and make important career choices. This is the transformation we are looking for in phase three. Even if you have lost weight and found better health already, the change is just beginning.

If, as you go into this final phase, you're struggling to keep up with your new habits, I recommend you try one (or more!) of these:

1. **Commit to a meditation practice.** This can be as simple as sitting with your eyes closed for twenty minutes before bedtime. Or maybe you leave your phone at your desk and use your afternoon walk as a movement meditation. No matter what you choose, a daily meditation practice lets you notice your thoughts without acting on them. By doing this, you learn that you are not your thoughts. This practice helps you learn that you are not any of the impulses that you have throughout the day. A mindfulness practice can do much more than strengthen your willpower; it can change the way you interact with other people all day long. You may find that your relationships improve with meditation and that your moods settle down.
2. **When something stressful happens, close your eyes and check in with your body.** Are you feeling queasy? Is your neck tense? Notice the physical effects of stress in the moment. Pay attention to these physical reactions, breathe into any distress, and allow yourself to relax. Do this before you reach for your cell phone to check email or send a text. You want to replace these urges with an even more powerful habit of listening to the body and checking in with your thoughts. Most of my patients start out unable to articulate why they live the way they do. It's just a routine they adopted over the years. At this point, you want to be connecting the dots between what you do that makes you happy and what makes you unhappy. Over time you will see that

you are holding your stress in one part of your body—the lower back, the stomach, or the neck. If you notice this, you will gain important information about how you live and which parts of your life are problematic. This is an important step toward setting new goals for yourself at work and at home. You can improve parts of your life that don't have anything to do with your diet and exercise routine.

3. **Ask yourself how it feels to exercise.** Turn your attention away from the notion of getting it over with. Instead, look for an activity that you can enjoy enough to look forward to. In those first few minutes of every workout, you may feel sluggish, but that good, energizing prana can give you a natural high that you can look forward to if you let yourself enjoy it. In this phase, I want you to find and focus on those activities you really enjoy, because those are the ones you'll stick with.
4. **Use the ginger drink (page 127) to help you combat food cravings.** Many times people use this drink in phase one to help themselves reduce their portion sizes and to abstain from unhealthy foods. You can go back to this drink now if you are feeling the return of food cravings.

As you settle into phase three, your day will look like this:

6:00 a.m.—Up and out of bed.

6:15 a.m.—Vigorous exercise. You still need just twenty to thirty minutes per day, but adding interval training will help you clear the channels and improve your fitness.

6:45 a.m.—Add in a five-minute meditation. Use the guide for sitting meditation on page 47. You can work up to a fifteen-minute meditation if you like. This will set your emotional equilibrium for the day.

7:00 a.m.—Eat a light breakfast.

8:30 a.m.—Finish with coffee or tea and start drinking water or herbal tea.

10:00 a.m.—Get up and move around a little, or get outside for natural light.

12:00 p.m.—Eat a healthy and substantial lunch.

12:30 p.m.—Get out for a little natural light.

1:00 p.m.—Set your intention to avoid snacks. Plan to drink two to three cups of water or herbal tea in the afternoon.

3:00 p.m.—Get up and move around a little. Get outside for natural light if you can.

6:30 p.m.—Eat a light dinner.

8:30 p.m.—Joyfully turn off your electronics and begin this phase of the day with a short meditation. You can do this for as little as five minutes or as much as fifteen minutes. This will do wonders for your sleep and for your emotional clarity the next day.

10:30 p.m.—Lights out for an effortless, deep, restorative sleep.

Did You Remember To:

- Try gentle fasting? Pick one day out of the week to give up dinner. This is an amazing technique for discovering the difference between fake hunger and real hunger. If you have a slower metabolism, you may have already tried giving up dinner once in a while to see how you feel when you go eighteen or more hours without food. This is actually an interesting way to test whether your hunger in the evening is real, too. You know you will be a little bit hungry at dinnertime, and you may be surprised at how quickly the hunger goes away after the dinner hour.
- Ask yourself three questions: *What did I do for my body today? What did I do for my mind today? What did I do for my spirit today?* It's easy to let the days slip by while you are working and putting out fires. But you need to pay attention to how you nourish your body with experiences, how you support your mind with new ideas, and how you uplift your spirit with meditation and reflection. The answers to these three questions will tell you how you live your life from day to day.

Herbs for Phase Three

There are several herbs that can help you clear your mind and strengthen your resolve, which is a big part of the transformative phase.

Ashwagandha. This herb is sometimes called "the strength of the stallion" because of its healing properties and its effect on the immune system. It helps to build muscles, particularly if you are engaging in intense exercise. It also calms and clears the mind. You can get this in powdered form and stir it into warm water or tea, or you can get it in pill form and take it in the morning or at night.

Brahmi. This is an adaptogen, meaning it helps the body adapt to new stressors. This herb supports the brain and can enhance memory, learning, and brain function. People say it helps them to think more clearly, especially when their lives are extremely busy. Yet it is best known for its ability to help reduce stress. It may also reduce inflammation throughout the body, which is another good thing to have going for you when you are releasing toxins in this phase of the diet plan. You can get tablets in 250-milligram doses and take it in the morning.

Shatavari. This traditional herb has been used for centuries as a tonic for women's reproductive health. It has a building and nourishing quality that helps ground you and balance your hormones. It provides natural support for estrogen production and can combat fatigue, meaning that it helps regulate cycles. Women can also take this when they are approaching menopause because it helps ease this transition. You can take a 500 milligram tablet in the morning, or stir ¼ to ½ teaspoon of the powder into warm milk or water.

I want to say that it can take seven to ten days to fully settle into this phase, but in reality, this phase is ongoing. As you deepen your meditation practice and fully connect to your body and your life goals, this phase will continue to change. Transformation is inherently a process, not a destination. During some weeks and months, this will be easy to maintain, while in others you may feel new challenges. By the time you've established a mind-body connection, you will find all sorts of changes happening in your life. You may be reaching for new challenges at work. You may find that your relationships improve. Or you may find that you have more confidence. Many of my patients who get to this point report that their friends are asking them why they look so good. They glow not just with good health but with good feeling. And that's my goal for you, too.

By setting and sticking to a healthy routine, you will become effortlessly in tune with your body and its needs. And in doing so, I hope you will become aware of time in a new way, not as a marker of how much you have accomplished or how much you still have to do, but rather as a series of opportunities. Morning doesn't have to be an exercise of dragging yourself out of bed and dashing to the office. Instead, it's a time to give yourself necessary energy, fuel, and contemplation. The noon hour is a time to stop what you are doing and nourish your body with a filling meal, a little bit of movement, and light. Most important, the evening will be a time to reconnect with yourself as you prepare to rest. I've said a lot about rest and rejuvenation in this book because it's what is missing most in the modern schedule. We are rushing around so much and so pointlessly that we forget how to stop and breathe and ask ourselves how we feel and to notice how fortunate we truly are. Think about the last time you were in an elevator: Did you push the "close door" button? Did you notice how worn down it was? It's not enough that the elevator is performing the miracle of lifting you many stories up in the air in just a few seconds. You feel the need to be annoyed that the door hasn't closed as soon as you are inside.

Ayurveda is the antidote to this ingrained impatience because it puts emphasis on rest and reflection. Ayurvedic scholars have long noted that the quality of your activity depends on the quality of your rest. And this has been confirmed by chronobiologists who note that the body operates in a master cycle every day, and that this cycle encompasses both activity and necessary rest. Both are crucial to good health. Ayurveda also urges awareness of your body's relationship to the larger world. It asks you to consider that what's going on outside in nature is also happening inside your body. These realms are intimately connected.

Ayurveda reminds you to be aware of how your body has individual needs. You don't need the exact same diet or exercise routine as your spouse or best friend. You have to know yourself and know your own needs. By checking in with your body routinely, you will glean wisdom about how to live well. In a way, Ayurveda is the original personalized medicine. By following the concepts in this book, you can keep yourself healthy not only every day, but in every season of the year and in every season of your life. And that's what I wish for you—effortless good health in every season.

Acknowledgments

This book would not have been possible without my clients and students who have brought me into their lives and shared with me their challenges and personal journeys. They have given me tremendous insights about the modern schedule and the tenacity of technology's hold on all of us, and they have shown real courage in changing their own lives. This book has given me an important understanding of how the right schedule can help you find purpose in life. As someone who moved from rural India to the busy corridor of Silicon Valley, I too struggled with balance in my life. Like me, my patients were working to juggle the demands of a busy job with the ideal of optimum health. Together, we tapped into the timeless Ayurvedic wisdom and found the answers, which became the inspiration for this book.

I also want to thank Amanda Annis at Trident Media whose endless enthusiasm for this project was evident from the first moment. She was a guiding light who kept telling me that, yes, this is an important book. Yes, people will need to read it. Very fortunately for me, Karen Rinaldi at HarperWave agreed and from our very first conversation, I knew I was in good hands.

This book would not have been possible without writer Michelle Seaton who brought intense curiosity about how Ayurvedic wisdom and modern science might come together. She has a razor-sharp ability to decode this complicated subject and distill it on the page, and she did so with hard work and creativity—in true vata-pitta style. Michelle enthusiastically adjusted her schedule and stayed healthy throughout the writing of this book despite competing personal and professional obligations.

My brilliant editor, Hannah Robinson, worked doggedly to make sure we had the right mix of science, personal stories, and Ayurvedic wisdom. Through several drafts, she cheerfully offered advice on everything from individual sentences to the structure of the book. And she knows the importance of staying on schedule. All authors need help with that. The whole team at HarperWave, including Yelena Nesbit, Penny Makras, Lydia Weaver, Leah Carlson-Stanisic, Adalis Martinez, and Erica Bahrenburg have worked so hard to make this a beautiful and engaging book. It has been a privilege to work with them all.

I am forever grateful for those who have formed the foundation of the wellness movement throughout the world. Without their teachings, nothing in this book would make sense to readers. His Holiness Maharishi Mahesh Yogi, the founder of the Transcendental Meditation Movement, gave me insights on higher states of consciousness, and also helped me understand the true nature of Vedic Science in its modern application. My deepest gratitude goes to Deepak Chopra, MD, for his inspiration, his vision, and his friendship. The Chopra Center is a leading light in mind-body medicine; and through his workshops and writings, Deepak has laid the groundwork for modern Ayurvedic medicine. He gave wholehearted support to this project from its earliest inception and has been an incredibly generous mentor for many years. I feel blessed to have learned from the very best.

I also want to thank my Ayurvedic colleagues who have worked tirelessly to make Ayurveda accessible to the western world. This includes my colleagues at Maharishi University who continue to inspire me. These are the beacons of Vedic wisdom and its integrative approach.

My staff at Ayurvedic Healing has been a solid rock throughout this process. They cheerfully adapt to my changing schedules and international travel while helping me to support the patients and clients who need our help.

Finally, and foremost, I must thank my wife, Manisha, who has been a quiet but dynamic force in my life. She held everything together through many years of travel and new endeavors. She accepted me as I am and quietly changed the fabric of my life to make it abundant, beautiful, and happy. She is a gifted Ayurvedic practitioner and author in her own right, and she has always been my first reader, the person who provides crucial insights on everything I write. Together we have two blessings in our children, Manas and Sanika, who have been firsthand reviewers of this material and its application. They are both vibrant, loving, and accomplished adults. Their unconditional love is my greatest gift.

Notes

Chapter 1: It's Not You; It's Your Schedule

1. Laura K. Fonken, et al., "Dim Light at Night Disrupts Molecular Circadian Rhythms and Affects Metabolism," *Journal of Biological Rhythms* 28.4 (2013): 262–71, accessed July 11, 2017, doi: 10.1177/0748730413493862.

2. Christoph A. Thaiss, et al., "Transkingdom Control of Microbiota Diurnal Oscillations Promotes Metabolic Homeostasis," *Cell* 159, no. 3: 514–29.

3. K. Kiser, "Father Time," *Minn Med* 88, no. 11 (2005): 26.

Chapter 2: Using Your Body's Internal Clock

1. M. Garaulet, P. Gómez-Abellán, J. J. Alburquerque-Béjar, Y.-C. Lee, J. M. Ordovás, and F. A. Scheer, "Timing of food intake predicts weight loss effectiveness," *International Journal of Obesity* 37, no. 4 (2013): 604–11, doi: 10.1038/ijo.2012.229.

2. A. D. Calvin, R. E. Carter, T. Adachi, et al. "Effects of Experimental Sleep Restriction on Caloric Intake and Activity Energy Expenditure," *Chest* 144, no. 1 (2013): 79–86, doi: 10.1378/chest.12–2829.

3. K. Van Proeyen, K. Szlufcik, H. Nielens, K. Pelgrim, L. Deldicque, M. Hesselink, P. P. Van Veldhoven, and P. Hespel, "Training in the fasted state improves glucose tolerance during fat-rich diet," *The Journal of Physiology* 588, no. 21 (2010): 4289–302, doi: 10.1113/jphysiol.2010.196493.

4. T. W. Puetz, P. J. O'Connor, and R. K. Dishman, "Effects of chronic exercise on feelings of energy and fatigue: a quantitative synthesis," *Psychological Bulletin* 132, no. 6 (2006): 866–76.

Chapter 3: Listening to Your Body

1. H. D. Critchley, S. Wiens, P. Rotshtein, A. Ohman, and R. J. Dolan, "Neural systems supporting interoceptive awareness," *Nature Neuroscience* 7 (2004): 189–95.

2. Barnaby D. Dunn, Iolanta Stefanovitch, Davy Evans, Clare Oliver, Amy Hawkins, and Tim Dalgleish, "Can you feel the beat? Interoceptive awareness is an interactive function of anxiety- and depression-specific symptom dimensions," *Behaviour Research and Therapy* 48, no. 11 (2010): 1133–38.

3. C. Price and K. Smith-DiJulio, "Interoceptive Awareness is Important for Relapse Prevention: Perceptions of Women Who Received Mindful Body Awareness in Substance Use Disorder Treatment," *Journal of Addictions Nursing* 27, no. 1 (2016): 32–38, doi: 10.1097/JAN.0000000000000109.

4. S. N. Garland, E. S. Zhou, B. D. Gonzalez, and N. Rodriguez, "The Quest for Mindful Sleep: A Critical Synthesis of the Impact of Mindfulness-Based Interventions for Insomnia," *Current Sleep Medicine Reports* 2, no. 3 (2016): 142–51, doi: 10.1007/s40675-016-0050-3.

5. C. E. Koch, B. Leinweber, B. C. Drengberg, C. Blaum, and H. Oster, "Interaction between circadian rhythms and stress," *Neurobiology of Stress* 6, no. 57 (2017): 57–67, doi: 10.1016/j.ynstr.2016.09.001.

Chapter 4: Sleep Is the Miracle Drug

1. T. Ronnenberg, et al., "Social Jetlag and Obesity," *Current Biology* (May 22, 2012): 939–43.

2. G. Potter, et al., "Circadian Rhythm and Sleep Disruption: Causes, Metabolic Consequences and Countermeasures," *Endocrine Reviews* 37, no. 6 (2016): 584–608, doi: 10.1210/er.2016-1083.

3. K. Spiegel, R. Leproult, and E. Van Cauter, "Impact of sleep debt on metabolic and endocrine function," *The Lancet* 354, no. 9188 (1999): 1435–39.

4. T. Roenneberg, A. Wirz-Justice, and M. Merrow, "Life between clocks: daily temporal patterns of human chronotypes," *Journal of Biological Rhythms* 18, no. 1 (February 2003): 80–90.

5. T. Ronnenberg, et al., "Social Jetlag and Obesity," *Current Biology* 22 (May 22, 2012): 939–43.

6. M. Boubekri, I. N. Cheung, K. J. Reid, C.-H. Wang, and P. C. Zee, "Impact of Windows and Daylight Exposure on Overall Health and Sleep Quality of Office Workers: A Case-Control Pilot Study," *Journal of Clinical Sleep Medicine* 10, no. 6 (2014): 603–11, doi: 10.5664/jcsm.3780.

7. A.-M. Chang, D. Aeschbach, J. F. Duffy, and C. A. Czeisler, "Evening use of light-emitting eReaders negatively affects sleep, circadian timing, and next-morning alertness," *Proceedings of the National Academy of Sciences* 112, no. 4 (2015): 1232–37, doi: 10.1073/pnas.1418490112.

8. J. W. Pennebaker and S. K. Beall, "Confronting a traumatic event: toward an understanding of inhibition and disease," *Journal of Abnormal Psychology* 95, no. 3 (1986): 274–81.

9. K. P. Wright, A. W. McHill, B. R. Birks, B. R. Griffin, T. Rusterholz, and E. D. Chinoy, "Entrainment of the Human Circadian Clock to the Natural Light-Dark Cycle," *Current Biology* 23, no. 16 (2013): 1554–58, doi: 10.1016/j.cub.2013.06.039.

10. Ellen R. Stothard, et al., "Circadian Entrainment to the Natural Light-Dark Cycle across Seasons and the Weekend," *Current Biology* 27, no. 4: 508–13.

Chapter 6: You Are When You Eat

1. S. Gill and S. Panda, "A smartphone app reveals erratic diurnal eating patterns in humans that can be modulated for health benefits," *Cell Metabolism* 22, no. 5 (2015): 789–98, doi: 10.1016/j.cmet.2015.09.005.

2. S. E. la Fleur, et al., "A daily rhythm in glucose tolerance: a role for the suprachiasmatic nucleus," *Diabetes* 50, no. 6 (2001): 1237–43.

3. S. Bo, et al., "Consuming more of daily caloric intake at dinner predisposes to obesity. A 6-year population-based prospective cohort study," *PLoS One* (September 24, 2014), doi: 10.1371/journal.pone.0108467.

4. D. Jakubowicz, et al., "High caloric intake at breakfast vs. dinner differentially influences weight loss of overweight and obese women," *Obesity* 21, no. 12 (2013): 2504–12, doi: 10.1002/oby.20460.

5. D. Jakubowicz, et al., "Meal timing and composition influence ghrelin levels, appetite scores and weight loss maintenance in overweight and obese adults," *Steroids* 77, no. 4 (March 10, 2012): 323–31, doi: 10.10.1016/j.steroids.2011.12.006.

6. C. Mu, et al., "Gut microbiota: the brain's peacekeeper," *Frontiers in Microbiology* 7, no. 345, doi: 10.3389/fmicb.2016.0345.

7. W. Z. Lu, et al., "Melatonin improves bowel symptoms in female patients with irritable bowel syndrome: a double-blind placebo-controlled study," *Alimentary Pharmacology & Therapeutics* 22, no. 927 (2005): 927–34.

8. S. Gill and S. Panda, "A smartphone app reveals erratic diurnal eating patterns in humans that can be modulated for health benefits," *Cell Metabolism* 22, no. 5 (2015): 789–98, doi: 10.1016/j.cmet.2015.09.005.

Chapter 8: The Right Exercise at the Right Time

1. T. W. Puetz, S. S. Flowers, and P. J. O'Connor, "A Randomized Controlled Trial of the Effect of Aerobic Exercise Training on Feelings of Energy

and Fatigue in Sedentary Young Adults with Persistent Fatigue," *Psychotherapy and Psychosomatics* 77 (2008): 167–74.

2. Y. Yamanaka, et al., "Effects of physical exercise on human circadian rhythms," *Sleep and Biological Rhythms* 4: 199–206, doi: 10.1111/j.1479-8425.2006.00234.x.

3. T. Miyazaki, et al., "Phase-advance shifts of human circadian pacemaker are accelerated by daytime physical exercise," *American Journal of Physiology-Regulatory, Integrative and Comparative Physiology* 281, no. 1 (July 1, 2001): R197–R205.

4. S. S. Tworoger, et al., "Effects of a yearlong moderate-intensity exercise and a stretching intervention on sleep quality in postmenopausal women," *Sleep* 26, no. 7 (November 1, 2003): 830–36.

5. Clara Bien Peek, et al., "Circadian Clock Interaction with HIF1α Mediates Oxygenic Metabolism and Anaerobic Glycolysis in Skeletal Muscle," *Cell Metabolism* 25, no. 1: 86–92.

6. E. Ulf, et al., "Does physical activity attenuate, or even eliminate, the detrimental association of sitting time with mortality? A harmonised meta-analysis of data from more than 1 million men and women," *The Lancet* 388, no. 10051 (2016): 1302–10, doi: 10.1016/S0140–6736(16)30370–1.

7. R. Mads, et al., "Body fat loss and compensatory mechanisms in response to different doses of aerobic exercise—a randomized controlled trial in overweight, sedentary males," *American Journal of Physiology-Regulatory, Integrative and Comparative Physiology* (August 1, 2012), doi: 10.1152/ajpregu.00141.2012.

Chapter 10: Your Body Through the Seasons

1. P. J. Brennan, G. Greenberg, W. E. Miall, and S. G. Thompson, "Seasonal variation in arterial blood pressure," *British Medical Journal* (Clinical Research Education) 285, no. 6346: 919–23.

2. D. J. Gordon, et al., "Seasonal cholesterol cycles: the Lipid Research Clinics Coronary Primary Prevention Trial placebo group," *Circulation* 76, no. 6: 1224–31.

3. R. Manfredini and F. Manfredini, et al., "Chronobiology of Vascular Disorders: a 'Seasonal' Link between Arterial and Venous Thrombotic Diseases?," *Journal of Coagulation Disorders* (January 2010).

4. J. M. de Castro, "Seasonal rhythms of human nutrient intake and meal pattern," *Physiology & Behavior* 50, no. 1: 243–48.

5. G. W. Lambert, C. Reid, D. M. Kaye, G. L. Jennings, and M. D. Esler, "Effect of sunlight and season on serotonin turnover in the brain," *The Lancet* 360, no. 9348 (2002): 1840–42.

6. C. N. Karson, K. F. Berman, J. Kleinman, and F. Karoum, "Seasonal variation in human central dopamine activity," *Psychiatry Research* 11, no. 2 (1984): 111–17.

7. R. D. Levitan, "The chronobiology and neurobiology of winter seasonal affective disorder," *Dialogues in Clinical Neuroscience* 9, no. 3 (2007): 315–24.

8. C. Meyer, V. Muto, M. Jaspar, et al., "Seasonality in human cognitive brain responses," *Proceedings of the National Academy of Sciences of the United States of America* 113, no. 11 (2016): 3066–71, doi: 10.1073/pnas.1518129113.

9. X. C. Dopico, M. Evangelou, R. C. Ferreira, et al., "Widespread seasonal gene expression reveals annual differences in human immunity and physiology," *Nature Communications* 6, no. 7000, doi: 10.1038/ncomms8000.

About the Author

Suhas Kshirsagar, BAMS, MD (Ayurveda) is a world-renowned Ayurvedic physician and educator from India, the director of the Ayurvedic Healing and Integrative Wellness Clinic in Northern California, and the author of *The Hot Belly Diet.* He holds a BA in Ayurvedic medicine and completed a three-year residency as an MD (doctorate in Ayurvedic internal medicine) with a gold medal at the prestigious Pune University. He is an advisor and consultant at the Chopra Center and a faculty member at several Ayurvedic institutions.

https://www.ayurvedichealing.net/
http://www.drsuhas.com/